Book of Extremes

California Central Coast: Courtesy of Jim White

Ted G. Lewis

Book of Extremes

Why the 21st Century Isn't Like the 20th Century

Copernicus Books is a brand of Springer

Ted G. Lewis
Technology Assessment Group
Monterey, CA
USA

ISBN 978-3-319-06925-8   ISBN 978-3-319-06926-5 (eBook)
DOI 10.1007/978-3-319-06926-5
Springer Cham Heidelberg New York Dordrecht London

Library of Congress Control Number: 2014939051

© Springer International Publishing Switzerland 2014
This work is subject to copyright. All rights are reserved by the Publisher, whether the whole or part of the material is concerned, specifically the rights of translation, reprinting, reuse of illustrations, recitation, broadcasting, reproduction on microfilms or in any other physical way, and transmission or information storage and retrieval, electronic adaptation, computer software, or by similar or dissimilar methodology now known or hereafter developed. Exempted from this legal reservation are brief excerpts in connection with reviews or scholarly analysis or material supplied specifically for the purpose of being entered and executed on a computer system, for exclusive use by the purchaser of the work. Duplication of this publication or parts thereof is permitted only under the provisions of the Copyright Law of the Publisher's location, in its current version, and permission for use must always be obtained from Springer. Permissions for use may be obtained through RightsLink at the Copyright Clearance Center. Violations are liable to prosecution under the respective Copyright Law.
The use of general descriptive names, registered names, trademarks, service marks, etc. in this publication does not imply, even in the absence of a specific statement, that such names are exempt from the relevant protective laws and regulations and therefore free for general use.
While the advice and information in this book are believed to be true and accurate at the date of publication, neither the authors nor the editors nor the publisher can accept any legal responsibility for any errors or omissions that may be made. The publisher makes no warranty, express or implied, with respect to the material contained herein.

Printed on acid-free paper

Copernicus Books is a brand of Springer
Springer is part of Springer Science+Business Media (www.springer.com)
Copernicus Books
Springer Science+Business Media
233 Spring Street
New York, NY 10013
www.springer.com

Preface

Even though the twenty-first century is still young, it is already marred by the dot-com crash, terrorism, financial system collapse, war, unsettling climate change, rise of new viruses—both animal and cyber—and an evolving socio-political shift caused by lightening speed advances in technology. Compared with the twentieth century, the first decade of the twenty-first century was as eventful and significant as the last 50 years of the previous century. Why?

We live in an age of rapid-fire change because over the past 200 years the comparatively straightforward Industrial Revolution has morphed into an era of nonlinear change punctuated with tipping points. The machinery of the current century is a collection of interconnected complex, rather than smooth-running, systems. Gradual and linear change no longer happens. Instead, "progress" moves in bursts—fits-and-starts marked by waves of unimaginable flashes, sparks, booms, bubbles, shocks, extremes, bombs, and leaps. Half probabilistic and half nonlinear deterministic, the twenty-first century is defined by intersecting long-tailed distributions, rather than independent and isolated Normal distributions. Episodic phenomena behave erratically and have no average or expected values.

Social, economic, physical, and cyber systems operate near their tipping points—perched on the edge of chaos, because they are optimized for maximum utility. Modern society has wringed out all surge capacity and backup reserves. After 60 years of stretching resources and budgets, governments have reached their limits. Traditional problem solving methods no longer work. Old ways of understanding our world no longer explain the current observed reality. Systems at all levels are out of bounds most of the time and increasingly collapse.

We need a new way to think about this new open, transparent, disruptive, and long-tailed world. We need tools for understanding the complexity of everyday things. Unfortunately, few of us understand complexity theory and nonlinear cause-and-effect, and even fewer of us are able to comprehend the implications of collapses, transformations, and revolutions currently taking place. Why did the global economy collapse? What are the principles underlying punctuated complexity? How can we begin to understand the nonlinear dynamics of global climate change, political upheaval, economic extremes, and the inevitable collapse and disaster confronting our society?

This sequel to Bak's Sand Pile, the author's earlier book, continues to explore society at its boundaries. From bursts of waves at the Asilomar Beach, to economic collapses, globalization of a tilted globe, and flashmobs precipitated by the Internet, Book of Extremes examines society's nonlinear outer limits.

Finding answers to these questions was the task I set for myself when I wrote *Bak's Sand Pile* in 2011. *Bak's Sand Pile* borrowed ideas from complexity theory, while this book applies nonlinear mathematics, catastrophe theory, big data, social network analysis, and biology to go a step further. This book proposes theories to explain extremes—why punctuated reality is bursty; why systems that have worked for decades suddenly fail; and why the wellbeing of entire nations is no longer in the hands of their leaders.

I begin with a gentle introduction to "wave theory"—the basis of punctuated reality. The theory explains social movements like the Arab Spring as well as online social media movements like Occupy Wall Street. Then I show how networks enhance nonlinear bursts—everything is connected to everything else—and hence a relatively small perturbation in one part of the space-time continuum ripples through the ether and impacts other parts. Unlikely events are made more likely by connected events. And connected events lead to conditional probabilities and the Bayesian theory underlying predictive analytics. [How we can forecast the future].

I have borrowed many ideas from biology and applied them to other fields such as economics. One of the most powerful ideas is the *Paradox of Enrichment*, which explains bubbles, shocks, and disruptions in financial markets. When combined with network science and nonlinear chaos theory, the unexpected crashes and disruptions in the global economy begin to make sense. For example, in Shocks, I show how comparative advantage leads to a highly interconnected global supply chain, which in turn determines the fortunes of entire nations.

Is reality completely random and unpredictable? Are we perched on the edge of disaster? Our understanding of mega-events such as the formation of social-political movements online leading to political instability in the Middle East, collapse of supply chains following the Fukushima nuclear disaster, global meltdown due to the 2008 financial crisis, impending ecological disasters from climate change, and plain old risk from terrorism and pandemics begins to crystallize as these mathematical concepts are placed in the context of social, political, and economic reality. This is perhaps the main contribution of this book—the application of rigorous methods to explain seemingly unexplainable events.

Book of Extremes is a work-in-progress, but it is a beginning.

March 2014   Ted G. Lewis

Contents

1	**Waves**		1
	1.1	A Day at the Beach	1
	1.2	Life Has a Long Tail	4
	1.3	The Fractal Structure of Reality	5
	1.4	Accidental Tourists	7
	1.5	Levy Walks	8
	1.6	Forecasting Accidents	9
	1.7	Why Are There So Many Long Tails?	11
	1.8	Everything Is Connected	11
	1.9	Casting Lots with Fate	12
	1.10	Sir Galton and Reverend Watson	14
	1.11	One More Thing	17
	Reference		19
2	**Flashes**		21
	2.1	Boulevardiering Is a Verb	21
	2.2	Drop Out, Tune In, and Join a Flashmob	22
	2.3	Flashmobs Are Punctuated Events	23
	2.4	Occupy Wall Street	25
	2.5	Internet Animali	27
	2.6	Mob Power: The Hub	31
	2.7	Put Out the Fire	33
	2.8	Forest Fires	33
	References		34
3	**Sparks**		35
	3.1	A Blip in the Night	35
	3.2	Knowledge Is Power, but Knowledge Is not Big Data	36
	3.3	Wizard of OS	38
	3.4	Complex Crowd Sense-Making	38
	3.5	21st Century Sense-Making	40
	3.6	Astroturfing	41
	3.7	An Old Technology	42
	3.8	Bayes' Network	43

3.9	Google's Car	45
3.10	The Sentient Network	46
3.11	Gamification	47
3.12	Human Stigmergy	48
3.13	Internet Intelligence	48
3.14	Going to Extremes	49
3.15	The Self-Referential Web	49
	References	50

4 Booms ... 51
4.1	Not So Natural Monopolies	51
4.2	Gause's Law	53
4.3	Self-Organization	54
4.4	Diffusion or Emergence?	56
4.5	Internet Hubs	57
4.6	Monopolize Twitter	59
4.7	Dunbar's Number	63
4.8	The Internet Is a Cauliflower	65
	References	67

5 Bubbles ... 69
5.1	Financial Surfer	69
5.2	Worthless Homes	70
5.3	The Paradox of Enrichment	72
5.4	Housing State-Space	73
5.5	US Economic Carrying Capacity	74
5.6	The Commons	76
5.7	Simple but Complex Ecosystems	77
5.8	The Road to Oblivion	79
5.9	A Moral Hazard	82
5.10	The Dutch Disease	83
	References	85

6 Shocks ... 87
6.1	The Richest Economist in History	87
6.2	Comparative Advantage	88
6.3	Extreme Economics	90
6.4	GDP: A Measure of Fitness	90
6.5	Chaos Theory	92
6.6	Master of Chaos	94
6.7	Tequilas and the Asian Flu	96
6.8	World Trade Web	96
6.9	The Raging Contagion Experiment	100
6.10	Securing the Supply Chain	101

	6.11	I Want My MTV	102
	6.12	Ports of Importance	103
	6.13	Shocking Results	104
	References		104
7	**Xtremes**		**105**
	7.1	The Big One	105
	7.2	The End of the World: Again	106
	7.3	The Carrington Event	108
	7.4	Black Bodies	110
	7.5	Models of Global Warming	112
	7.6	Anomalies in the Anomalies	113
	7.7	Global Warming Isn't Linear	114
	7.8	Global Warming is Chaotic	115
	7.9	The Lightning Rod	117
	7.10	Green's Forcing Function	118
	7.11	Consequences	120
	7.12	Solar Forcing	120
	7.13	Other Forcings	122
	7.14	Pineapple Express	122
	References		129
8	**Bombs**		**131**
	8.1	The Malthusian Catastrophe	131
	8.2	Old, Poor, Crowded, and Oppressed	133
	8.3	The Wealth Effect	134
	8.4	The World Is Tilted	135
	8.5	Distributed Manufacturing	136
	8.6	Concentrated Financial Control	137
	8.7	The World Is a Bow Tie	138
	8.8	Concentration of Power Is a Network Effect	140
	8.9	Pareto Was Right of Course	143
	8.10	Econophysics: Wealth Is a Gas	144
	8.11	Redistribution	146
	8.12	Wealth Redistribution Isn't Natural	147
	References		148
9	**Leaps**		**149**
	9.1	Not Waiting for ET	149
	9.2	Levy Flights of Fancy	150
	9.3	X-Prize Leaps	151
	9.4	Leaps of Faith	153
	9.5	Un-constrain Your Constraints	153
	9.6	Like Prometheus Stealing Fire	155

	9.7	Disruptive Rootkit	157
	9.8	Fill a White Space	157
	9.9	Work in the Weeds	159
	9.10	Hack Your Culture	161
	9.11	Extreme Exceptions	162
	References		163
10	**Transitions**		**165**
	10.1	One in a Million	166
	10.2	This Is Your Life	167
	10.3	Out of Nothing Comes Something	168
	10.4	The Commons	170
	10.5	The Wealth of Nations	172
	10.6	Extreme Challenges	173
	10.7	The World May Be Flat, but It Is Also Tipped	174
	10.8	The Millennium Falcon	175
About the Author			177
Index			179

Waves 1

Abstract

Terrorist attacks, floods, nuclear power meltdowns, economic collapses, political disruptions and natural events like earthquakes come and go in waves—they are bursty. Why? The modern world differs from previous generations mainly due to vast network connections that replace independent random events with highly dependent conditional events. The 21st century is an age of lop-sided long-tailed probability distributions rather than the classical Normal distribution. Like episodic waves striking the beach, modern day events come and go in bursts—most are clustered together in time and space, but occasional bursts are separated in time, space, and consequence. This model of reality is called *punctuated reality*, because of the wave-like behavior of the complex systems found everywhere in modern life.

1.1 A Day at the Beach

Pacific Grove California is one of nature's most beautiful places. The world-renowned Monterey Bay Aquarium located at the end of David Avenue attracts millions of visitors to the jellyfish, seahorse, and sunfish exhibits. Down the street from the aquarium is Asilomar State Beach—a rustic wind-swept compound of Julia Morgan buildings where Silicon Valley mavens like Steve Jobs and Larry Ellison often met to decompress and rejuvenate the troops.

I trekked to the Asilomar beach on a sunny day in June to observe the waves as they crashed onto the rocky shore. Wave after wave crashed against a certain bolder I had picked out as a landmark. For the next 20 min I recorded the time intervals between subsequent waves using the timer on my *iPhone*. The first two waves were separated by 36 s; the next wave quickly arrived one second later; and the next two swept over the landmark bolder after a lapse of 11 and 14 s,

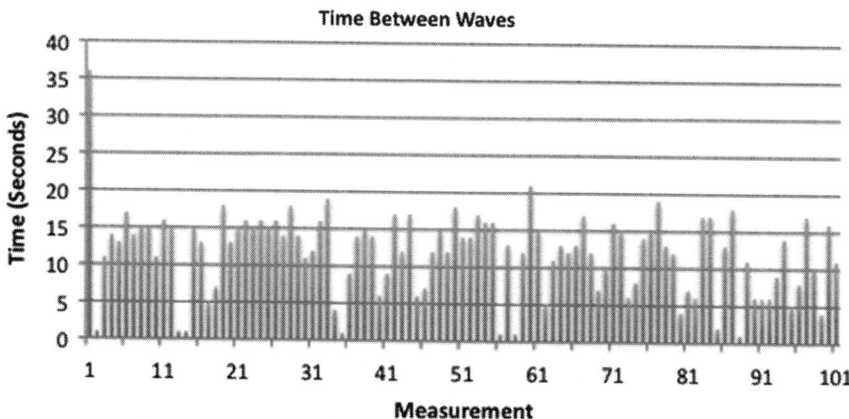

Fig. 1.1 Time intervals between waves appear to be random noise. Measurements were taken sequentially over a period of time

respectively. In 20 min I recorded over 100 time intervals on my notepad for analysis later on. Figure 1.1 shows my data—plotted as elapsed time *between waves* for the entire 20-min period. Notice any pattern?

I returned home and entered the time intervals into my computer and constructed the frequency plot shown in Fig. 1.2. The raw data in Fig. 1.1 looks like random noise but the processed data in Fig. 1.2 shows a distinct pattern. The distribution of time intervals plotted in Fig. 1.2 forms what is called a *long-tailed distribution*. When plotted as a histogram, the time intervals between waves no longer look like noise on a TV screen. Seemingly random wave motion turns out to be somewhat predictable.

If the pattern of wave action was truly random, its frequency distribution would form a symmetrical mound-shaped curve—a *Normal distribution*—with most measurements falling on either side of a representative value and others trailing off to either side. Random waves *should* form such a bell-shaped curve, so the long-tailed pattern came as a surprise. Even more surprising is the fact that long-tailed distributions like the one in Fig. 1.2 have no meaningful average value. While it is mathematically possible to calculate an average value and standard deviation from the data, neither value represents anything physically or practically meaningful, because long-tailed distributions extend to infinity. What is the average value over an infinite collection of points? That is, there is no such thing as an average wave interval! Every wave is considered an atypical *outlier*. [Wave intervals turn out to be an extreme valued statistic].

So, what does the lop-sided distribution tell us about waves? The distribution of Fig. 1.2 says that most waves reach the shore in closely timed intervals, while fewer and fewer waves washed ashore after relatively long lapses of time. Waves arrive in *bursts*—rather than rolling onto the shore in rhythmic patterns. Arrival times are irregular—almost chaotic—but with a distinct pattern as shown in Fig. 1.2. These simple Asilomar Beach waves contain a hidden order as revealed

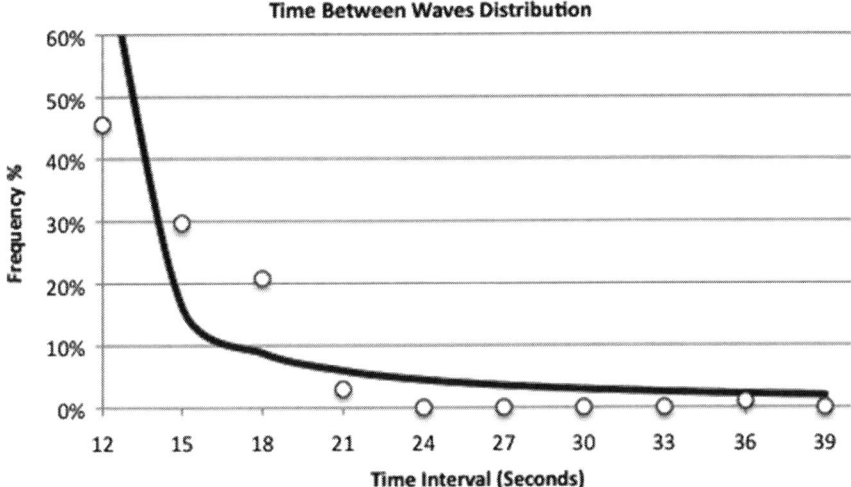

Fig. 1.2 Time intervals obey a long-tailed distribution. Dots are data points obtained from measurements in Fig. 1.1, and the solid line is a best-fit power law of the form, $f = t^{-q}$, where f is the frequency of elapsed times, t, and q is a constant that best-fits the data

by the long-tailed distribution. Waves are simple, and yet they are complex. They behave somewhat randomly, but also with some amount of predictability—specifically, we can expect them to be bursty or punctuated. Just when we think we know the pattern, it changes. Perhaps this is why waves mesmerize idle scientists with time to waste on Asilomar Beach.

I was unable to predict when the next wave would reach the bolder I had picked out as my landmark. Nor was I able to predict how big the wave would be. Bursts of incoming waves were followed by indefinite periods of inaction. Sometimes two or three crests arrived at nearly the same time. Other times, the sea was rather calm and tranquil. Instead of the smooth rhythmic motion I had come to expect from my high school science teacher (waves are caused by the rhythmic gravitational pull of the moon), I found that wave reality is rather *punctuated*. Even worse: it occurred to me that bursts and punctuations are metaphors for modernity.

Wave behavior is like many 21st century phenomenon—a combination of rhythmic and chaotic patterns. A systematic rhythm is established for a relatively long period of time only to be disrupted by an outlier—a miniature tsunami or longer-than-expected calmness. Sound familiar? Worldly events chug along at an orderly pace for months only to be disrupted by a burst of extreme events. For example, the Soviet Union collapses, setting into motion a series of social-political events that eventually die down. The world enjoyed a relatively peaceful and uneventful mid-to-late 1990s until the dotcom bubble burst, setting off another chain reaction of political and financial upsets. Fortunes are followed by reversal of fortunes. Bursts of political anxiety are followed by unusually tranquil periods of calm. And then the unexpected happens, without warning.

1.2 Life Has a Long Tail

Like the waves of change sweeping over our society as I write this book, nature is a combination of regular predictability and irregular unpredictability. But even nature's chaotic behaviors obey certain basic rules. In the case of ocean waves, the chaotic collisions between water and rock conceal a deeper pattern. The long-tail distribution of Fig. 1.2 models this combination of predictability and unpredictability—it is impossible to precisely predict the arrival time of the next wave, and yet the frequency of time intervals between subsequent waves isn't entirely erratic and unpredictable—it follows a well-defined distribution.

The predictable part of wave intervals is easy to understand. Most intervals are short, while only a few are long. The unpredictable part is more challenging. The precise number of seconds that pass between subsequent waves is somewhat random. Thus, wave intervals are both predictable and unpredictable. In the large, they obey a lop-sided long-tailed distribution as shown in Fig. 1.2, but in the small, it is impossible to predict exactly when the next wave will arrive.

Figure 1.2 was obtained from the raw data displayed in Fig. 1.1. However, instead of plotting the wave intervals sequentially, the frequency of intervals of equal size is plotted versus elapsed time interval. The horizontal axis of Fig. 1.2 is divided into bins of equal size and the number of wave intervals falling into each bin is tallied to obtain a histogram. Then, normalize the histogram by dividing each bin count by the total number of measurements. This process yields a frequency distribution as shown by the dots in Fig. 1.2.

The solid line shown in Fig. 1.2 is obtained by fitting a long-tailed power law to the binned data. As you can see, it is lop-sided—short wave intervals are more likely to occur than long intervals. Extremely long intervals are even more unlikely, although they are not too rare. This is the point: the distribution of elapsed time between wave intervals is biased. If it were not, an equal number of data points would fall on either side of an average value. Instead, to my surprise, most points fell on the left side of the frequency diagram. And, there is no average value separating short and long intervals.

Long-tailed distributions like the one shown in Fig. 1.2 indicate a degree of predictability even as there is an element of randomness. Short intervals are more likely than long intervals, and so a betting person would bet that the next wave interval is shorter, rather than longer. But, the wager is still a gamble, because an occasional long interval is still likely. Thus, Figs. 1.1 and 1.2 both contain an element of predictability as well as chance.

This is where my story begins: why is the modern world often governed by long-tailed distributions instead of the well-known Normal Distribution? Why do we observe so many events that are composed of both predictable and unpredictable elements? Is the world truly punctuated—a series of erratic bursts followed by relative calm? How can something as simple as tidal wave action be complex? And on a grander scale, how is it that society is both simple and complex at the same time?

1.2 Life Has a Long Tail

As it turns out, this is the first clue to how the 21st century operates: both small and large events exhibit properties of predictability and unpredictability. Some things can be anticipated with regularity, while other things seem to defy comprehension. The real world is defined by long periods of time where nothing very important happens, followed by a rapid series of events close together in time and space. And typically, the unpredictable bursts matter a lot, because they change the world. This is called *punctuated reality*.

1.3 The Fractal Structure of Reality

Initially I was worried about how to measure the time-between-waves, because some waves were small while others were big. Should I sort out the intervals according to how large or small they were? As it turns out, it doesn't really matter, unless of course, the wave is a tsunami! [In that case, I should run!]. Big waves and small waves illustrate another fact of modern life called *scale*. If I had flown over the beach in an airplane, I would have been able to count only large waves. If I had gazed into the ocean from a mere 6 inches, I would have been able to count very small waves. The aerial view obscures the small waves, while the 6-inch close-up view obscures the big waves. Scaling does not seem to matter, however.

Scale is another property of the punctuated 21st century—both big and small events are subject to the same long-tailed fingerprint. Aerial and close-up wave-counting experiments both produce long-tailed distributions, but at different scales. We say that wave intervals *scale*, because whether I measure the intervals from near or far, both measurements produce a long-tail distribution. This curious fact is profound, because it says something about the similarity of earth-shaking events versus insignificant events—earth-shaking events are a lot like everyday small events—only bigger. It is a matter of scale.

Long-tailed distributions like the one in Fig. 1.2 are called *scalable*, or *self-similar*, because they are *fractals*. In simple terms, a fractal looks the same at all scales. Take a magnifying glass to a long-tailed distribution like the ones shown in this book, and you get another long-tailed distribution! They all look identical regardless of how near or far away they are observed. Ocean waves and many worldly events such as changes in stock market indexes look similar at different scales—that is, they exhibit self-similarity. The underlying patterns are the same regardless of how far or close we observe them.[1]

This is the second lesson of the 21st century that makes it different from the 20th and earlier centuries. Many more events in modern life obey fractal or self-similar distributions than ever before in history. [This is a big claim that will take the rest of this book to justify]. Big waves (tsunamis) are just like small waves (Asilomar Beach), in terms of frequency of time intervals. The time scale (or size scale) may change, but the underlying phenomenon is the same. Big events mimic small events,

---

[1] Changes in a stock price obey a long-tailed distribution and so do wave intervals.

and vice versa. For example, big events like the Arab Spring in Tunisia and Egypt in 2011 mimic self-similar small events like Occupy Wall Street, and the American Tea Party movement. Even small *flashmob* events that pass with little notice are fractals—disturbances operating at a relatively small scale. Similarly, protests in China occur at all scales, but when aggregated into a frequency distribution, they obey a long-tailed distribution just like the waves in my experiment.

> Fractal nature of nature: Observations made at different scales often belie the same underlying structural dynamics—punctuated, self-similar, and long-*tailed*.[2]

Some say that history repeats itself, but I say history obeys fractals. In many cases, the fractal is a long-tailed distribution like the ones shown in this book. Thus, large-scale hurricanes, terrorist attacks, nuclear power meltdowns, financial collapses, and earthquakes are simply scaled-up small hurricanes, attacks, nuclear power hazards, financial disruptions, and earth tremors. Incidents like these happen all the time, but most of them go unnoticed (fall on the left end of the long-tailed distribution), while a small number of rare-sized events blow up in our face (and fall on the right end of the long-tailed distribution). Instead of repeating itself, history repeats a self-similar pattern at different scales.

The dimension of self-similar events may differ, however. If the frequency distribution of worldly events fits a power law as shown in Fig. 1.2, the exponent in the power law represents the *fractal dimension* of the phenomenon. A large fractal dimension, q, means the distribution is very lop-sided and short-tailed. A small fractal dimension means the distribution is relatively flat and long-tailed. For example, the frequency of earthquakes obeys a long-tailed distribution with fractal dimension near 1.5.[3]

Long-tailed fractal distributions tell us a lot about small and big events in the course of people's lives. Principally, the lesson is this: change is constantly bubbling under the surface of day-to-day reality. Most bubbles are ordinary. But, on rare occasions change is so large and disruptive that it takes our breath away and fascinates entire nations for a long period of time. In fact, major changes such as the terrorist attacks of 9/11 not only get our attention, they foment a period of *chaotic adaptation* or adjustment to a "new world order". Then, after a while life returns to a smaller scale—the so-called "ordinary life".

What would life be like if we could take note of the small constantly bubbling punctuations or surges happening in everyday life? What if we consciously monitored the stock fluctuations of a favorite company, the political fluctuations of national politicians and their policies, or the economic fluctuations of businesses around the world? Would it be possible to anticipate and prepare for the big shocks? If we knew the significance of small events, might we be able to anticipate and even exploit the inevitable big surges? Is it possible to anticipate the future?

---

[2] As I will show, the rules governing much of modernity are non-parametric, non-linear, scalable, conditional, and regulated largely by extreme statistics.

[3] Caneva and Smirnov (2004).

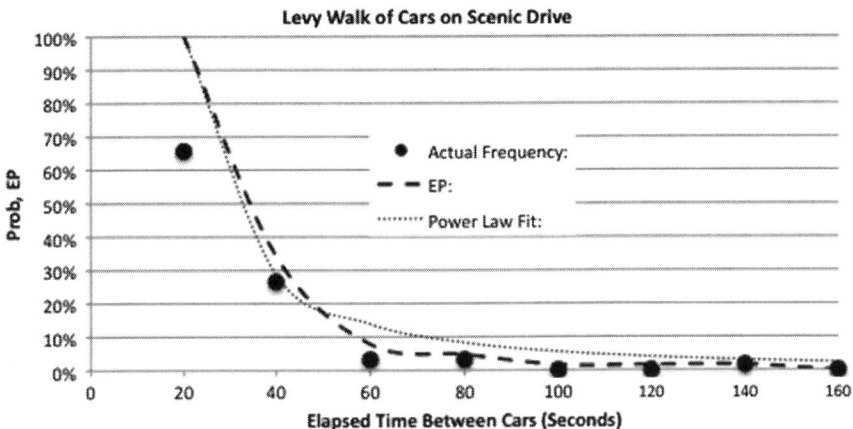

Fig. 1.3 Cars are like waves: they come and go in bursts. Actual counts turned into a frequency histogram are compared with the same data converted into an exceedance probability distribution along with its best-fit power law

1.4 Accidental Tourists

Ocean waves crashing against the rocks below me were not the only punctuated events happening while I was counting waves. Tourists driving their rental cars and riding their bicycles past my lookout point along Sunset Drive also fascinated me. So I turned my attention to counting the elapsed time between cars! Sure enough, cars passed by me in waves—sometimes in a cluster of 3 or 4 together, other times only one car at a time. In rare instances, the bursts were separated by relatively long intervals of time. But most of the time the elapsed time between cars was short. In fact, the distribution of time intervals were much like the long-tailed distributions shown in Fig. 1.2, see Fig. 1.3.

Figure 1.3 introduces a new measure of punctuated reality—the *exceedance probability distribution*. Like the frequency distributions of Fig. 1.2, the exceedance probability distribution of Fig. 1.3 is long-tailed. At each point along the horizontal axis, x, the exceedance probability, EP, represents the probability of an interval as big or bigger than x. For example, EP(40) is the probability that an interval of time equal to or greater than 40 s will occur. Think of EP as a worst-case measurement—it is the likelihood of intervals of size 40 or more seconds between cars passing by me as I count them.

Theoretically, the time separating random cars driving along any road should be a short-tailed distribution—a lop-sided mound called the *Poisson distribution*, which has an average time interval aligned with the mound's peak. If tourists really appear randomly out of thin air—as if by accident—the distribution of cars would not only look different than shown in Fig. 1.3, but it would have an average value just like the Normal distribution. But my measurements produced a distribution nearly identical to the one describing intermittent ocean waves. Surprisingly, the cars traveling along Sunset Drive are punctuated, too. They obey the same bursty rules of as the famous Asilomar beach waves.

Why? For one thing, tourists traveling along Pacific highways of California do not pop into sight purely by accident. They are influenced by forces beyond their control. Slower cars cause packs to form, an unscheduled deer-crossing or otter sighting can make a tourist pause, and local residents come and go at rather predictable times of the day. And then there is the occasional Butterfly Parade of costumed children that stops traffic for hours!

Accidental tourism isn't entirely accidental. Instead, people and events of historical note are like waves approaching the beach. They come and go in bursts. These are called *Levy walks* or *flights*—a term coined by Benoît Mandelbrot in honor of the French mathematician Paul Lévy (1886–1971). People, animals, and some contagious diseases move about the surface of the earth along paths that obey the long-tailed fractal shown in Fig. 1.3. That is, most displacements in space are close together, while a few are far apart. People, animals, and many events of great importance occur on the surface of the earth separated by distances distributed as a long-tailed curve.

Furthermore, Levy walks are characterized by long-tailed exceedance probabilities that model worst-case displacements. EP is a measure of extreme behavior—one of the observed properties emerging in the 21st century. We should replace the outmoded Normal distribution and its average value, with the non-parametric, lopsided, extreme-valued exceedance probability distribution, because it models reality better than the smooth, well-behaved Normal distribution.

1.5 Levy Walks

Is the shift from average to extreme-valued distributions an accident or is there a deeper meaning to these long-tailed phenomena? Are long-tailed fractals simply a mathematical trick that applies only to pathological cases like counting cars and waves?

As I sat next to the Pacific Ocean counting waves, my mind began working out an explanation for punctuated reality. Far across the Pacific, but not very far away in time, another wave had recently struck with deadly force. A tsunami created by an earthquake under the ocean northeast of the Japanese islands swept over the northeast coast of Japan, overwhelming the retaining wall protecting the nuclear reactor at Fukushima dai-ichi. [Reactor number one in Fukushima Prefecture]. A chain reaction that began in the Pacific Ocean propagated disaster through miles of Blue Ocean, seashore, retaining wall, and very thick cement walls into the heart of one of Japan's nuclear power plants. This was the immediate consequence of the Great *East Japan Earthquake* (also known as the *2011 Tohoku earthquake*).

The previous unexpectedly and relatively large nuclear power accident was 25 years earlier at Chernobyl, Ukraine. And the one at Three-Mile Island was 7 years before Chernobyl. Nuclear power plant accidents happen all the time, but we only hear about the worst ones. [Scale is the only difference between a big event and a small one]. Wikipedia lists 23 notable accidents since 1952. See Fig. 1.4. And the distribution of time intervals separating them obeys the same long-tailed distribution that waves and cars at Asilomar Beach obey. It seems that bursts of catastrophes in the space-time continuum are subject to the same self-similar rules!

The long-tailed data illustrated here are often plotted in log-log format in the scientific literature, because a truly self-similar distribution appears as a straight line in a log-log plot. Figure 1.4b is the same data as shown in Fig. 1.4a, but plotted on log-log coordinates. If the phenomenon under study is self-similar, observations should fall on the straight line. In other words, scientists use the log-log plot to determine if a certain phenomena like nuclear power plant incidents are fractals. The slope of the straight line in Fig. 1.4b is the fractal dimension of nuclear power plant accident fractal.

Figure 1.4 isn't a perfect fit, but it comes close to a textbook perfect long-tailed, self-similar, exceedance probability distribution as before. Is this a coincidence or a hint that something deeper is going on?

1.6 Forecasting Accidents

The fractal behavior of accidents like the Fukushima dai-ichi disaster gives scientists a tool for estimating what might happen in the future. Nobody can predict the future, but we can use the past to estimate the likelihood of a similar event in the future. This tool is illustrated by the projection of nuclear power incidents going forward, in Fig. 1.4c. The likelihood of a reportable incident rises to near certainty between now (2013) and 2018. [This prediction says nothing about the size of the predicted event, only that it will eventually happen].

The projection assumes nuclear power plant incidents—both small and large—are separated in time by Levy walks. That is, my forecast is based on the observation of past nuclear power plant incident Levy walks, and the assumption that the past is a predictor of the future. In mathematical terms, this is called an a posteriori estimate, because it is based on the past. Once we know the fractal dimension of the *a posteriori* curve, we can project the fractal forward in time.

How reliable is the forecast in Fig. 1.4c? If all of the nuclear power plant incidents since 1999 were removed from the data, and the exceedance distribution re-calculated, the a posteriori prediction would forecast the next significant event to occur with probability of 95 %, instead of 92 %. Specifically, if constructed in 1999, the probability distribution of Fig. 1.4c would have predicted a nuclear power event of some significance to occur by 2011 with probability of 95 %. Fukushima dai-ichi falls in line with such a prediction.

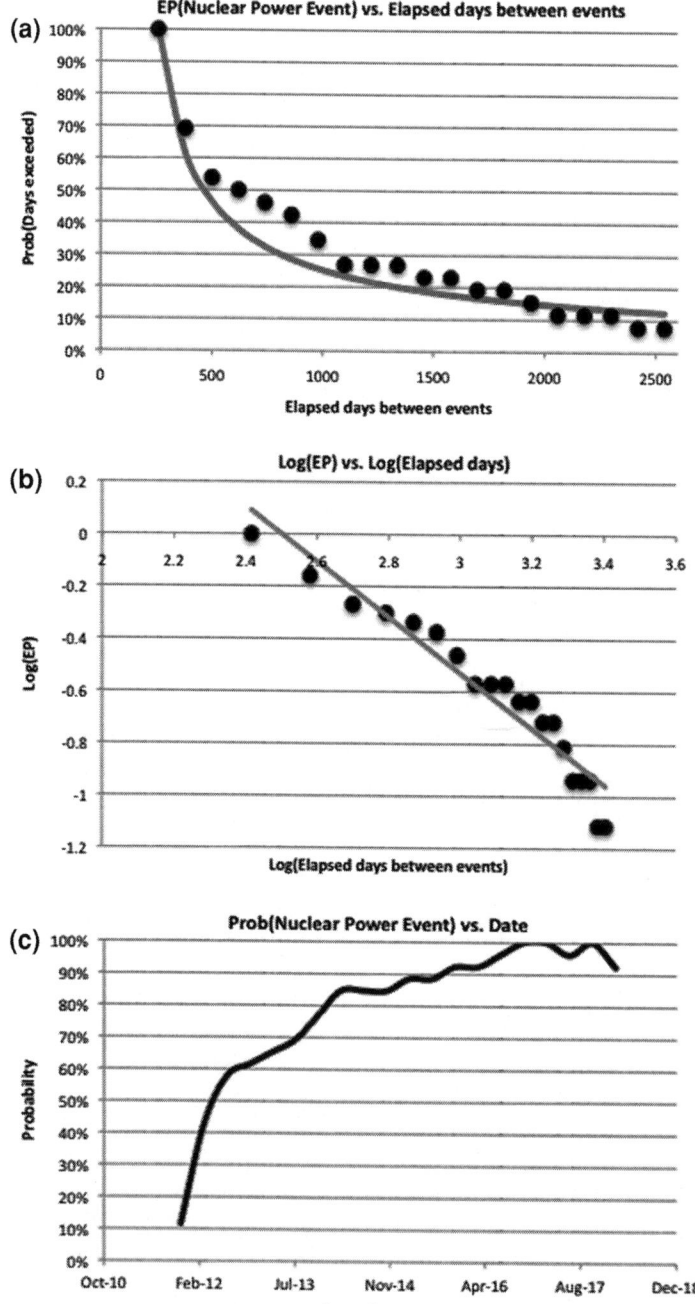

Fig. 1.4 Time intervals between nuclear power accidents surge just like waves on a beach. The data is presented in a log-log format in (**b**). Forecast of the next incident of any reportable size is shown as an exceedance probability estimate in (**c**). **a** Exceedance probability distribution of nuclear power incidents and a best-fit power law curve. **b** The data in (**a**) shown as a log-log plot. A perfect power law plots as a straight line in log-log space. **c** Forecasted date of next nuclear power event, based on past incidents and the exceedance distribution in (**a**)

Nuclear power plant incidents, terrorist attacks, hurricanes, floods, fires, and many other catastrophic events obey long-tailed distributions with fractal dimensions, which in turn can be used to forecast the likelihood of events that are similar to events that have happened in the past. This is the basis of forecasting used by insurance companies as well as authors of books on boom and doom. Why does this simple tool seem to work?

1.7   Why Are There So Many Long Tails?

From observations of waves, cars, nuclear power incidents, and others described throughout this book, one thing stands out: evidence mounts that they all obey a long-tailed fractal space-time distribution. This raises a deep question regarding disastrous incidents: "why is so much of the modern world governed by long-tailed fractals?" What is the underlying "cause and effect"?

Fractal versus Normal Distribution reality is the third major difference between the 20th and 21st centuries: The 20th century can be explained through the application of Normal Distributions. Not so in the 21st century. In this century, events obey fractals in all dimensions: size, time, and space (distances apart). One might say that modern events move through the space-time continuum according to the extreme statistics of Levy walks. Modernity tends to obscure these fractals, but they exist, none-the-less.

The reason: the 21st century is highly connected.

1.8   Everything Is Connected

Bursts aren't isolated events that affect just a few people, places, and things. They are part of constantly propagating chain reactions circling the globe. Like hundreds of domino chains collapsing all at once, waves, cars, power plants, power blackouts, terrorist attacks, airplane crashes, and Internet exploits are simultaneously and constantly spreading waves of large and small incidents. Most are small and inconsequential. Others are large enough to be notable. A very small number of toppling dominoes cause huge and threatening cataclysms. Ordinary everyday dominoes go unnoticed, while existential dominoes get all of the attention because they fall and threaten complete extinction—perhaps once every millennium.

The 9.0 Great East Japan Earthquake off the coast of Japan set in motion a series of events whose consequences magnified as it spread. Fifteen thousand people in Japan perished. Millions of other people around the world were also affected. The chain reaction triggered a major nuclear disaster of global proportions, damaged the supply chain of Toyota Motors and the hi-tech industry of the United States and Europe, not to mention a total disruption of Japanese life. The release of dangerous levels of radiation—detected as far away as Asilomar Beach,

California—was just the most obvious consequence. The damage to various supply chains and industries had far more serious consequences. At the time of this writing, the full effects of Fukushima's disaster have not been fully calculated.

The Fukushima dai-ichi incident dramatically illustrates another principle that will become more important in the 21st century: the world is *networked*. In a networked world, an incident at one point in the space-time continuum propagates to other points in the space-time continuum. And, if the incident is large enough, the impact is felt at many other "nodes" or points in space-time.

Fukushima dai-ichi is a node in a global network connecting all of us. One link from Fukushima dai-ichi extends to the supply chains that provide goods and services from/to Asia and the United States. Additionally, there are links to the economy, public health, political groups, and many other nodes in the global system. The global network radiating out from Fukushima dai-ichi is so big I cannot show all of it to you. [The World Trade Web is analyzed in *Shocks*—the chapter that analyzes nonlinear economics and the effect of financial shocks on the wealth of nations].

The impact of punctuated reality is not restricted to physical catastrophes. Similar domino effects exist in other sectors. For example, a debt crisis in Greece spread to Spain, Italy, Germany, England, and the United States. A villager in China contracted SARS and the epidemic killed people in 29 other countries in a matter of months. An uprising in Tunisia contaminated the surrounding nations and changed the policies of nations around the world—a reverberation still going on today.

Network connections are profound "vectors" because they provide pathways through space and time to other points in space and time. Waves of change pass through these vectors at alarming speed. More to my point, Network contagions are long-tailed incidents—they surge in time, forage in space, and occasionally cause big collapses. They obey the same fractal rules described above.

The connectedness of most everything to just about everything else has a major impact on modern society. But why do these obviously disparate social, political, economical, and physical systems behave in almost identical ways? Why does the long-tailed distribution keep cropping up every time we look deeper into the machinery of everyday events? The answer lies deep in probability theory. Which naturally brings me to *gambling*.

1.9 Casting Lots with Fate

Gambling with 6-sided cubes or "tossing dice" is over 5,000 years old. One would think something that old is so well understood by modern humans that we could move on to quantum theory or space travel. But tossing dice still fascinates and entertains, because games of chance add an element of uncertainty to life. In fact, until the 1650s, people thought the Gods of Chance controlled gambling outcomes—who won and who lost. But Blaise Pascal put all that to rest when he invented the mathematical theory of probability in 1654. From that enlightened moment on, casting lots has become a science, not a mystery.

1.9 Casting Lots with Fate

Table 1.1 All of the possible outcomes of rolling two dice and adding them together are shown as a table. The pips of each die are along vertical and horizontal borders, while the sums are shown at the intersections

Two dice	1	2	3	4	5	6
1	2	3	4	5	6	7
2	3	4	5	6	7	8
3	4	5	6	7	8	9
4	5	6	7	8	9	10
5	6	6	7	9	10	11
6	7	8	9	10	11	12

Pascal was a child prodigy. In addition to founding modern probability theory, he invented one of the first mechanical calculators and made important contributions to hydrodynamics. His breakthrough in probability theory was based on a simple, but elegant assumption: the outcome of a game of chance is easily calculated by dividing the number of *favorable* outcomes by all possible outcomes. He called this fraction between zero and one, a *probability*. Suddenly, the magic was gone, but the challenge of foretelling the future remained, because a probability only tells us the likelihood of an outcome—not its certainty. Pascal's brand of chance is called a priori, because it calculates the likelihood of a future event without any past. That is, a priori probability is pure math, not history.

So let us do the experiment that Pascal and others did over 350 years ago. Toss two dice and count the number of *pips* (black spots) appearing on the top of each cube. Table 1.1 lists all possible outcomes obtained by summing the number of pips on two dice. Each die can land upright and show 1, 2, 3, 4, 5, or 6 pips. Thus, there are 6 times 6 equals 36 possible outcomes. This is the number we divide into the number of favorable outcomes to get Pascal's probability fraction. For example, suppose the favorable outcome is a total of 11 pips. From Table 1.1 we see there are two ways to obtain a total of 11. Thus, the probability of obtaining an 11 is 2/36, or 5.5 %.

The likelihood of obtaining a score of 11 on any one toss does not guarantee an 11 every time the dice are tossed. Instead, it simply tells us what the chances are. Probability theory is the science of likelihoods, not certainties. Even with this caveat, we have also assumed the dice are fair. That is, Table 1.1 assumes each die will land on one of its six sides with equal probability. This assumption is rarely true in real life. [Hold that thought for now].

Now, suppose we change the problem slightly. Instead of tossing both dice at the same time, suppose we toss one die, first, place our bets, and then toss the second die after revealing the number of pips appearing on the first die. For example, suppose the first die turns up 4 pips. What is the probability that the total will reach 11, now? Consulting Table 1.1 and reading along row 4, it is clear that it is impossible to achieve a total of 11 even after the second die is tossed! Thus, the probability of achieving a score of 11 drops from 5.5 %, to zero!

We say the probability of an 11 is *conditional* on the value of the first die. The probability of achieving a score of 11 depends on the outcome of the first die. For example, if the first die is 4, the probability of getting a total of 11 is zero. If it is 6, the probability rises to 16.6 %. Now two events are linked together so that the outcome of one is dependent on the outcome of the other. The probability of an 11 is conditional on the score obtained on the first toss.

In the bursty world of the 21st century almost everything is connected to almost everything else. Thus, the likelihood of an event in one part of the world is no longer independent of events occurring elsewhere. Instead, the likelihood of an event is *conditional*. Thus, the probabilities must also be conditional.

What does *conditional probability* mean in practical terms? First, we must recognize that events are often sequential, so that the probability of an event is conditional on a previous and sequential event. Knowing that something has already happened is useful information. We say this foreknowledge reduces uncertainty, which is a kind of knowledge. Information becomes knowledge and knowledge becomes valuable as uncertainty is reduced. More important, knowledge can be exploited to reduce *risk*.

In the 21st century we have the technology to convert data and unstructured information into valuable knowledge using previously unheard of computational power. Some call this *big data*, because turning information into knowledge often requires processing massive amounts of data. Terabytes and petabytes of data can sometimes tell companies what consumers want, farmers how climate affects crops, and law enforcement agencies where the next crime will happen.

The 21st century is different from the 20th century because we can process information into knowledge and reduce uncertainty. This takes big data, but with the proliferation of the Internet, big data is becoming a commodity. In fact, big data is sometimes called the *new oil*, but without the Middle East as a controlling intermediary. Given a large enough database and a lot of computing cycles, it is now possible to perform a number of feats unimaginable to the 20th century mind. The new field of *predictive analytics* was invented to perform these feats on a daily basis.

The 21st century is conditional, because connected systems are interdependent, which means the likelihood of future events is also conditional. Events such as the financial meltdown of 2008, the Fukushima dai-ichi nuclear power catastrophe, and political upheaval in the Middle East are linked together and conditional. Long-tailed phenomenon like changes in stock prices, spread of pandemics, political unrest, and shortages in global supply chains are the manifestation of linked events, conditional probabilities, and our modern ability to reduce uncertainty through the application of predictive analytics. All of this takes big data and big computers to do the work. This is where the Internet comes in!

1.10 Sir Galton and Reverend Watson

Anyone that watches the British Broadcasting Corporation (BBC) knows that Victorians were prigs. One of the most priggish of prigs was Sir Francis Galton (1822–1911). He exceeded all expectations. Galton worried that aristocratic surnames might vanish from the rolling English hills unless something was done to

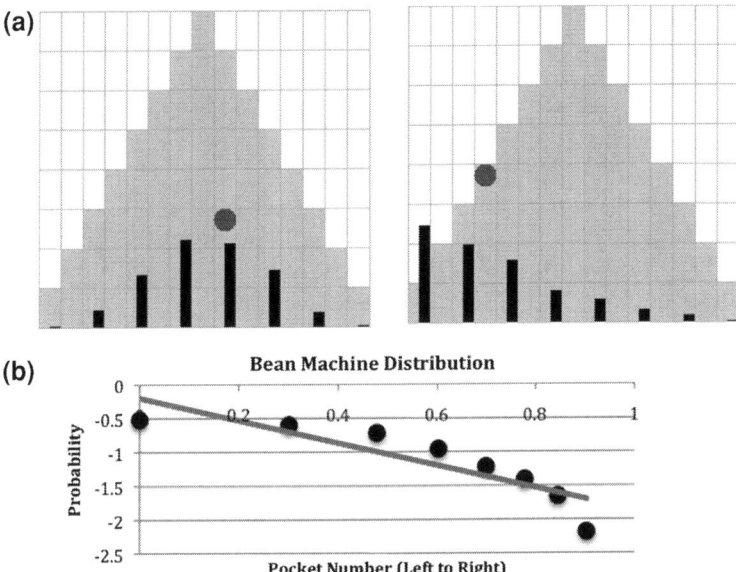

Fig. 1.5 a The Bean Machine in action. Independent probability on the *left*, and conditional and dependent probability on the *right*. Binomial distribution is shown on the *left* and long-tailed distribution on the *right*. **b** Log-log plot of the probability of a bean landing in a certain pocket

stop surname dilution. His concern for the longevity of aristocratic names led him to pose the question to a broader audience in a paper he wrote for the *Educational Times* of 1873. His paper stirred up further action, as one might expect from such a priggish act.

Reverend Henry William Watson quickly replied with a mathematical answer. In 1874 the Galton-Watson duo published the solution to the inconsequential name problem in a paper titled, "On the probability of extinction of families". And sure enough, the two gentlemen showed that "good names" were almost certain to die out over time, especially if the nobility indiscriminately produced female children. After all, the best names are handed down through sons of nobility, not the daughters.

Galton and Watson are perhaps better remembered for the Galton-Watson *bean machine*. If you have visited any science museum of repute you perhaps observed the bean machine in action, see Fig. 1.5. Here is how it works. Beans are released at the top of the machine and allowed to fall willy-nilly through a series of left-right intersections. By chance each bean goes into either the right-hand or left-hand tube as it falls toward the bottom of the machine. If everything is equal, the beans end up in a pile shaped like a mound, as shown by the frequency distribution in the left-most diagram of Fig. 1.5a. In fact, if equality is assumed the mound-shaped pile obeys the famous *binomial distribution*, which is a discrete version of the famous Normal Distribution. More beans fall near the middle of the bean machine than at either extreme. When this happens, we say the beans tend to *regress to the mean*.

Regression to the mean is a fundamental concept in statistics. Just about everything we learned in school assumes it. But is it reality? As it turns out, regression to the mean, binomial distributions, and independent events are mostly a myth. In reality, much of the world is governed by conditional probabilities and lop-sided or long-tailed distributions. The Galton-Watson bean machine models an ideal world that doesn't exist. In most cases, the probability of beans going in one direction versus another depends on certain pre-conditions. And these pre-conditions occur in real time, right before each bean has to make a decision to turn left or right.

What happens if the bean machine processes beans on a conditional basis? Suppose we rig the bean machine so that the probability of a bean moving to the left is dependent on where the bean has been before. That is, suppose the likelihood of moving left at each opportunity *increases* as the bean moves left. The more left turns a bean makes, the more likely it is to make another left turn. Therefore, the probability of taking a left branch at any intersection in the bean machine depends on whether or not the bean has previously taken left turns. Each bean makes a decision on which way to go, based on its history.

The path taken by beans is no longer a random walk. Instead, it is a *Levy walk*. As the bean falls from one intersection to the next, it accumulates a history of conditional probabilities. Once a bean turns left, it is more likely to turn left again. This dependency produces a long-tailed distribution as shown on the right in Fig. 1.5a. And, to make the data more presentable to the scientific mind, Fig. 1.5b confirms the fractal hypothesis by showing the conditional bean machine obeys a fractal long-tailed distribution instead of a Normal Distribution.

The conditional probability bean machine is a clue to the puzzle of why many events governing the 21st century are long-tailed fractals, and here it is: in the 21st century, events are linked, which also means their probabilities of happening are linked. The likelihood of a power outage in your neighborhood increases if the power grid in the next county has failed. The likelihood of contracting a disease increases if your neighbors have it. And the likelihood of being stuck in a traffic jam also increases if the people ahead of you are jammed. Life is conditional, not independent. Long-tailed fractals are born out of conditional probabilities.

This is a big concept. In the 20th century we could get away with assuming major events occurred independent of other events, because there were only a few or no connections between them. In the 21st century, it is highly likely that a major event on the other side of the globe is linked somehow with systems and events on this side of the globe. So when something happens in say, New York City, its ramifications are felt in London, Beijing, and Sao Paulo. The beans not only fall according to the path taken up to the current point or intersection, but the beans end up in a state that is described by a long-tailed distribution. And long-tailed distributions favor small events over large events. In terms of global connectivity, long-tailed events are rare, powerful, and significant, simply because they are linked to a series of other events and systems that preceded them. One might say that the more links you have with the outside world, the more likely an event from afar will impact life in your backyard.

In the 21st century, it is foolish to assume unexpected events happen out of the blue. Rather, the Fukushima dai-ichi nuclear disaster was linked to the tsunami, which was linked to the earthquake, which in turn was linked to tectonic plate movement in the earth's crust. And who knows what caused the plate movement? But, what we now know is that Levy flights, time intervals between waves and cars, catastrophes like Fukushima dai-ichi, and many other "earth-shaking events" of significance have one deep characteristic in common: they obey lop-sided long-tailed fractal distributions because of conditional probabilities.

But conditional dependence is still an oversimplification of punctuated reality.

1.11 One More Thing

There is one more factor to consider before abandoning this line of reasoning. In addition to being conditional, modern events are governed by extreme values. Extreme events follow different rules than the normal events that typically obey the Normal distribution and regress to the mean. Long-tailed distributions are non-parametric, meaning they have no mean value. Long-tailed fractals follow the rules established by the field of *extreme statistics*, meaning they have no defining parameters like mean value or standard deviation.

Normal statistics describe events that are pulled toward expected values. Extreme statistics describes events that are pulled toward extreme values. The most useful extreme value statistics are those describing the likelihood of minimum, maximum, or the product of a connected series of events. Figure 1.6 shows plots of both minimum values and products of values selected randomly. Note how the product of probabilities is a more extreme fractal than the distribution for maximum values.

The plots of Fig. 1.6 were obtained by computer simulation. Three random numbers were repeatedly generated and in Fig. 1.6a, multiplied together to produce the distribution. In Fig. 1.6b, the three random numbers were compared, and only the smallest one kept. Its distribution is derived from counting only the smallest of the three numbers. The three-number sample is repeated thousands of times to generate the histograms shown here.

So what? These simple simulations show that the statistics of extreme values behave just like the fractals observed in the real world. These artificially produced distributions are identical, in most cases, to what we observe in catastrophic hurricanes, floods, wars, terrorist attacks, and financial collapses. That is, extreme value statistics provides an adequate model of punctuated reality.

Figure 1.6a forms an ideal long-tailed distribution by multiplying conditional probabilities together. Similarly, the probability of another Fukushima dai-ichi-like event is the product of probabilities: the probability of a major earthquake times the probability of a tsunami, times the probability of swamping the protective seawall, times the probability of reactor meltdown. The probability of a series of conditional events propagating like a series of dominoes across the Pacific

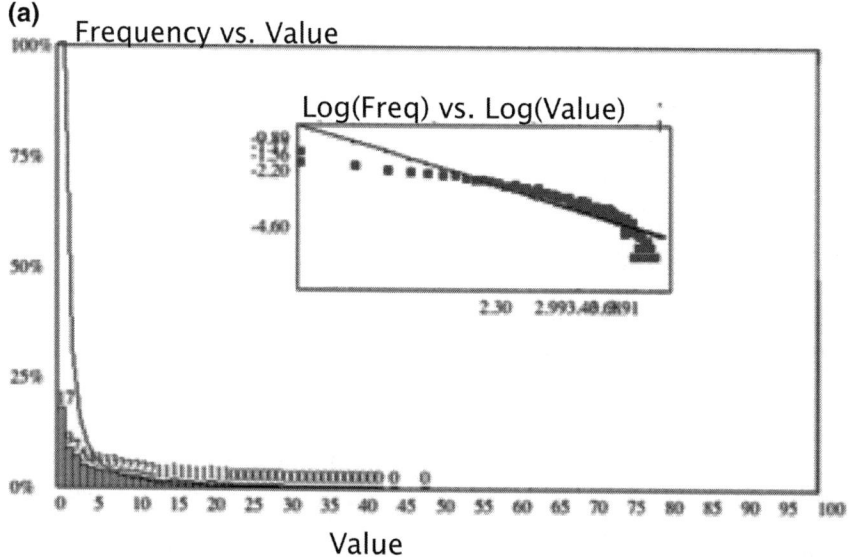

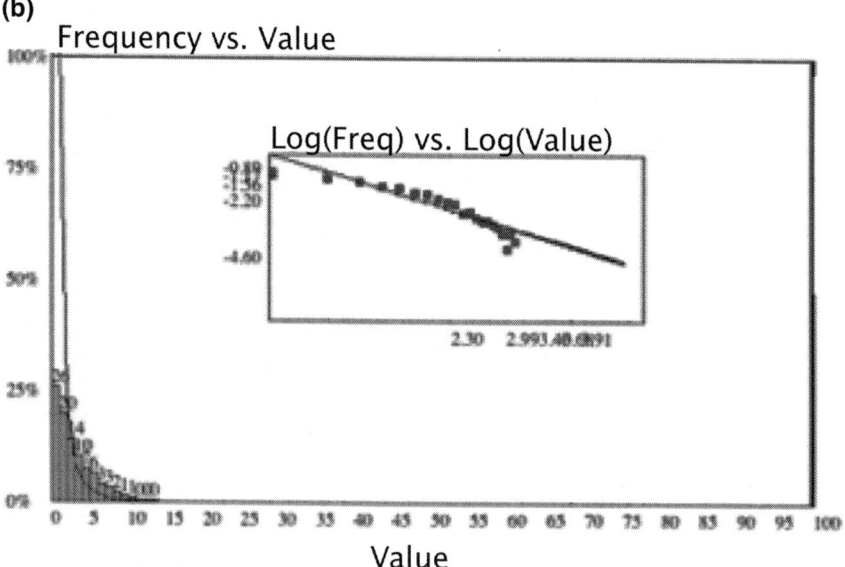

Fig. 1.6 Comparison of the statistics of extreme values shows that extreme values obey long-tailed fractal distributions. The product of probabilities is more extreme (longer-tailed) than the probability of minimum values. **a** Product of 3 random values produces a long-tailed distribution. **b** Minimum of 3 random values produces a long-tailed distribution

1.11 One More Thing

Ocean to render massive damage on the Japanese island obeys the same probability distribution of Fig. 1.6a, because they are both produced by multiplying conditional probabilities together.

The bottom line is that the highly punctuated, connected, conditional, extreme-valued 21st century is governed by long-tailed fractals instead of symmetrical error functions like the Normal Distribution with its averages and standard deviations. Fractals obey extreme value statistics, and the claim of this book is that the 21st century does too. In summary:

> Reality is punctuated: it obeys long-tailed Levy flights in space and time with fractal dimensions. Long-tailed Levy flights are the manifestation of conditional probabilities. Conditional probabilities are the result of a highly connected world. Connectivity is the underlying mechanism of modern complexity, which produces long-tailed distributions in place of Normal distributions.

Reference

Caneva, Alexander, and Vladimir Smirnov, "Using the Fractal Dimension of Earthquake Distributions and the Slope of the Recurrence Curve to Forecast Earthquakes in Columbia", Earth Sci. Res. J. Vol. 8, No. 1 (Dec. 2004) : 3–9.

Flashes 2

Abstract

Because punctuated reality is episodic—waves come and go—the social structure of modern life is also episodic. Social groups form spontaneously and often without obvious reasons. Flashmobs ravage innocent bystanders, online raves spring up and just as quickly die out, protesters flair-up on Wall Street and then disappear. Why? These bursts occur because of the increasingly connected super hive we call the Internet. More importantly, they rise and fall because of how society is wired together. Ever-tightening self-organization and the resulting hub-like structure of society magnifies the impact and spreads these events like a wildfire. If you want to control the Internet animali, seek out and control the center.

2.1 Boulevardiering Is a Verb

Boulevardiering is one of the most endearing customs of urban Italians—parading up and down major thoroughfares of Rome and other Italian cities in one's finest clothing. It is largely an extrovert's sport played by social animals with an abundance of self-confidence. Boulevardiering regularly breaks out among the Neapolitan natives near the Castel Nuovo off Via Nuovo Marina Boulevard or most anywhere the stylishly dressed Italians happen to go in the cool evening after siesta and before dinner at 9 pm. Italians love to be spontaneous, but with style. They fondly call these spontaneous exhibitionists, *"Boulevard Animali"*—parading animals.

Boulevardiering has spread to other, more uptight countries like the USA. Southern Methodist University students in Dallas Texas have been doing it for years. Typically the SMU Mustangs parade around the campus prior to a big game against the rebel Black Bears of the University of Mississippi. It started in 2000 as a kind of extemporaneous celebration in honor of the new Gerald J. Ford Stadium. SMU needed something bigger and better than the Black Bear tailgating parties in Mississippi. So they turned "boulevard" into a verb—an act of one-upmanship over the University of Mississippi. It must have worked, because SMU students have been boulevardiering ever since.

Boulevardiering holds no surprises. As the evening wind dies down and the Italian sky turns auburn, people turn out gradually at first, and then in droves. And when dinnertime arrives the crowd fades just as orderly and smoothly as it gathered. Whether parading around in one's fashionable attire in Italy or baseball cap and war paint in Texas, the ritual is a predicable one—smooth and rhythmic as one would expect from a civilized and sophisticated people.

Boulevardiering is nothing at all like flashing.

2.2 Drop Out, Tune In, and Join a Flashmob

In contrast to boulevardiering, another kind of human animali spontaneously organize throughout the world for no apparent reason. Like gathering storm clouds that eventually turn into a torrent of rain, these animali unleash a torrent of abuse and destruction on the people and buildings around them. In some countries, like the USA, the assembly of youth looking for a thrill has become a major problem for law enforcement. These flashes of humanity are called *Flashmobs*. And, they can kill.

On September 29, 2002, in Milwaukee, Wisconsin, 16 young people suddenly appeared and beat a man to death after throwing an egg at him. On November 23, 2007 one hundred people ransacked a convenience store for no apparent reason. Seven people brutally attacked a man in a parking garage in Columbia, Missouri on June 6, 2009. One hundred and twenty juveniles tore up a Wal-Mart store in Cleveland Heights, Ohio. Teens accosted and assaulted innocent pedestrians in Silver Springs, Maryland, and 950 people went on a rampage in Philadelphia, the City of Brotherly Love, in 2010. During a short decade from 2002 to 2012, over 150 incidents were reported at http://violentflashmobs.com.

Wikipedia defines a *flashmob* as, "a group of people who assemble suddenly in a public place, perform an unusual and seemingly pointless act for a brief time, then disperse, often for the purposes of entertainment, satire, and artistic expression. Flashmobs are typically organized via telecommunications, social media, or viral emails".[1] Spontaneous email and message texting are favorites of the Internet cognoscenti. Unfortunately, many of these spontaneous combustions are neither entertaining, satirical or artistic. They are downright malicious.

Flashmob techniques have recently taken on deeper social and political meaning because of their power to organize and control Internet animali. For example, flashmob techniques have been used to foment revolutions

---

[1] http://en.wikipedia.org/wiki/Flashmob

(*Arab Spring*), instigate political change (*Occupy Wall Street*), and encourage social change (American Tea Party, Gay Rights Parades, and Raves). Flashmobs have become a tool for the prankster, marketer, and political activist. These planned and unplanned bursts appear to build up over time, reach a peak, and then die out, only to repeat at a later date. They are the 21st century equivalent of rebels without a cause.

What sets modern flashmobs apart from old-fashioned rebellion is the speed and agility made possible by the Internet. Mobsters can quickly organize and deploy a flashmob using *Facebook, Twitter*, cell phones, or other social media. These 21st century technologies extend the reach of the disrupters, crossing borders, time zones, cultures, and social classes. In the modern world, boulevardiers are being replaced by *Internet animali*—people glued to their mobile telephone or tablet, ready and willing to be deployed on an instant's notice. But unlike peaceful boulevardiers, flashmobs can turn uncivilized.

The speed and reach of Internet technology make word-of-mouth, organized meetings, and planned social interactions obsolete. Today's town hall meeting is an extreme event created, organized, and fed by *Internet animali* prowling the Internet. Mobs spontaneously appeared and instigated pillow fights with Toronto shoppers, attacked their math teacher in Irvington, New Jersey, and ran a truck over a 49-year old man in Jackson, Mississippi. Are these simply accidents, or carefully planned military-like exercises perpetrated by individuals exercising their powers of Internet persuasion?

2.3 Flashmobs Are Punctuated Events

Flashmobs and other technology-amplified movements among the animali fit the rules governing 21st century culture like an intellectual glove. First, these surges obey long-tailed distributions in size, time, and space, just like the waves lapping at my feet on Asilomar Beach. Flashmobs recorded from 2002–2010 in North America moved through space and time like a foraging animal or Manhattan shopper. They are *Levy flights*. See Fig. 2.1 through 2.3. Big events are self-similar to small events in size, elapsed time between events, and distance from the last event.

Figure 2.1 through 2.4 show *exceedence probability* plots of 150 flashmob incidents listed at http://violentflashmobs.com/. Recall that the exceedence probability is simply a probability curve where the vertical axis represents the likelihood of an incident *equal to or greater than the x-axis*. Exceedence probability curves measure the probability of an event exceeding some level, as indicated along the x-axis.

We use the exceedence probability to estimate the probability of flashmob events of a certain size, elapsed time between events, displacements, and future incidents in size, location, and time. From Fig. 2.1 you can see that most flashmobs are small, but a few rare ones are large. There is no such thing as an average

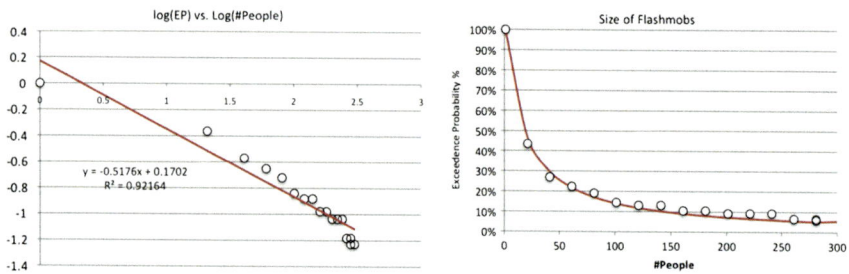

Fig. 2.1 Most flashmobs are *small*, but some are *large*. Size obeys a *long-tailed* exceedence probability distribution

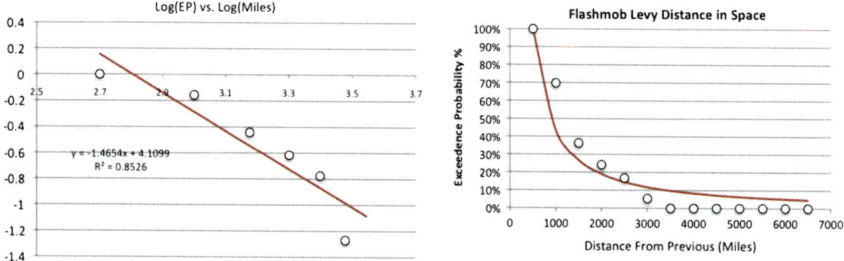

Fig. 2.2 Flashmobs are separated in space by distances that obey a *long-tailed* exceedence probability distribution

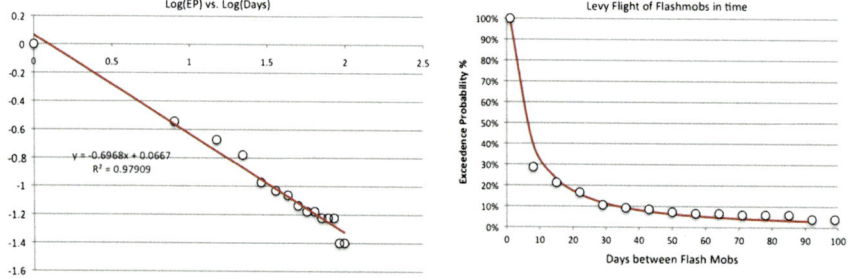

Fig. 2.3 Flashmobs also obey a *long-tailed* exceedence probability distribution in time: flashmobs are mostly clustered together in time, but a few are separated by long periods of quiet

uprising, rave, or flashmob incident. Instead, there are many insignificant uprisings, raves, and flashmobs happening all the time, but once in a while really significant flashes happen. Some flashmobs unleash such a torrent that they topple entire governments. These are the so-called *black swans* [Lewis 2011].

2.3 Flashmobs Are Punctuated Events

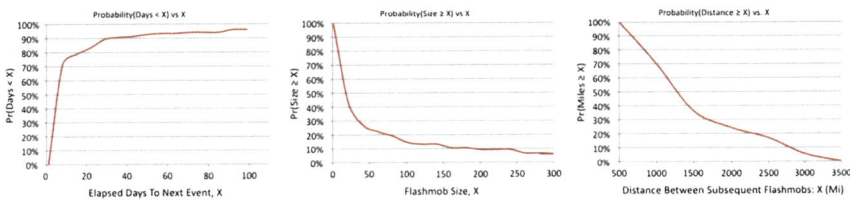

Fig. 2.4 Probabilities of a flashmob in the size-space-time continuum obey long-tailed distributions and follow Levy Flights

Flashmob distributions are also *fractals* because of their self-similarity. Big events mimic small events, and vice versa. A magnified slice of the curve in Fig. 2.1 is shaped just like the entire curve, signifying that large mobs are magnified small mobs. [We also say self-similarity is a form of *scalability*, because changing the scale of the plot does not change its shape]. The slope of the exceedence curve plotted on log-log paper equals the fractal dimension of the phenomenon. In fact the length of the curve's tail is quantified in terms of the slope or fractal dimension of the curve. A long-tailed exceedence curve says big events are rare, but still more likely to happen than expected. The tail is longer and fatter for low fractal dimensions, and shorter for high fractal dimensions.

Flashmobs follow *Levy flights* around town and sometimes across the globe. Spatial Levy flights mean that mobs occur on the surface of the earth separated by distances distributed as a long-tailed curve. Figure 2.2 illustrates this for the flashmobs recorded in North America. Flashmobs are more likely to recur near the same location as previous mobs, and less likely to occur far away. Occasionally, the mob jumps relatively long distances with probability determined by the tail end of the curve.

Due to the global Internet, these bursts of social mayhem are the byproduct of connected tremors propagating from one place to another along fault lines in society. Sometimes the connections are obvious, as they were with the Arab Spring revolution. The uprising in Tunisia and the tremors that followed in Egypt were connected through a variety of Internet-based social networks. Other times the connection is subtle, as was the case with the Occupy Wall Street movement in 2011. Flashmobs occur more often where there is a high concentration of self-organized activists on the verge of disruption. Sometimes this critical point is reached on its own, while other times it is contrived, as was the case with the Occupy Wall Street movement.

2.4 Occupy Wall Street

Kallie Lasn and Micah White run *Adbusters*—a magazine devoted to making money by bashing other people that make money. Lasn and White are avowed anti-consumerists, anti-capitalist, and anti-government anarchists. Lasn has been quoted as predicting "a dark age" following the collapse of capitalism, because capitalists simply consume too much.² In fact, according to Lasn, "killing capitalism" is inevitable because there are too many people wasting too many limited resources for the consumer-obsessed system to remain viable. To hasten the end of modern society, Lasn and White have devoted their lives to launching discontent and watching how it spreads throughout society. Their brand of flashing is more traditional, but it uses the new technology of the Internet to spread the word.

Adbusters is a compelling read. The articles are well researched and well written. A 2012 article on the Internet culture boasted that the Internet would lead to, "a slow erosion of our humanness and our humanity". In another article, an *Adbuster* writer ranted, "the relentless mass marketing of cool has tainted ... [our] behavior" [Schwartz 2011, p. 28]. Making money from criticizing capitalism is apparently quite profitable, because *Adbusters* has been around since 1989.

Adbusters is a slick publication. It has great graphics, thought-provoking articles, and a large following. In fact, the magazine is so persuasive that when Lasn called upon his 70,000 readers to protest Wall Street mischief in November 2011, thousands of people showed up. Lasn single-handedly sparked the *Occupy Wall Street* (OWS) movement, and almost as quickly lost control of it.

Lasn has the smarts to instigate a revolution. With a degree in mathematics and years of experience doing market research for the capitalist companies he now despises, he has the skill to "throw ideas into the culture that then have a life of their own" [Schwartz 2011, p. 30]. He treats ideas like a cold virus—when he sneezes his readers catch his cold. Lasn and White cooked up the idea of Occupy Wall Street in early July 2011. Then an email campaign followed an ad in *Adbusters*. Twitter.com and Reddit.com became infected, and the people behind OWS began their boulevardiering. Justine Tunney in Philadelphia put the notice on her RSS feed and registered OccupyWallStreet.org with the Internet's Assigned Names authority, IANA.

Meanwhile, White contacted *New Yorkers Against Budget Cuts* and convinced them to join the boulevardiering anarchists. He went to work on fragmented groups of discontents like the one proposing a *Robin Hood Tax* on the richest 1 %, *Irish Hunger Strike* advocates, and people against the poor treatment of American Indians. Soon, it was the 1 % versus the 99 %. The contagion even jumped the Atlantic Ocean. David Graeber, a London-based anarchy theorist, added flames to the spreading bonfire. An over-the-hill activist from the 1960s, Tom Hayden, recommended that the movement remain ambiguous in its goals for the time being. Apparently, the best revolutionaries are the ones nobody can understand.

Soon, the OWS idea spread over the country like swarming locusts. The signature of the mounting online flashmob appeared as a radar blip on Facebook. Figure 2.5 plots the exceedence distribution of Facebook postings versus the number of likes from various cities in the USA during the build up in late 2011.³

---

² Schwartz (2011).

³ http://www.collectivedisorder.com/occupytogether/latest

2.4 Occupy Wall Street

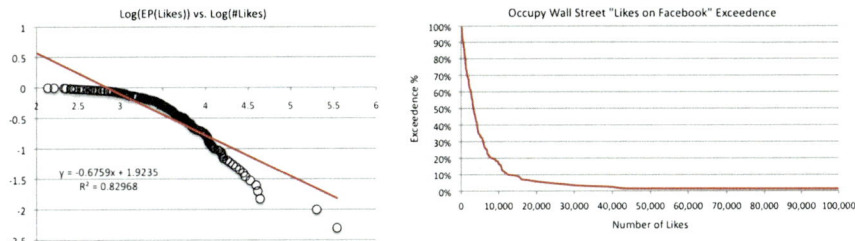

Fig. 2.5 Occupy Wall Street postings on Facebook ranged from a maximum in excess of 100,000 to a minimum of 93 during the period: October 5, 2011—November 29, 2011

And as expected, its exceedence probability curve is a long-tailed distribution, with an R^2 value of 82 %. It seems that discontented 99 %-ers behave a lot like the waves lapping the shore at Asilomar Beach. They left a trail of Levy Flight breadcrumbs on the Internet for us to study.

But then the OWS sunburst was over. Within a few months the world turned its attention to the Syrian civil war, the US Presidential elections, and the miserable state of the global economy. Like SARS forging a path out of China into nearby countries and then dying off in a matter of 6–8 months, the Adbuster's campaign no longer mattered much. A flash is simply that—a sudden burst followed by a relatively long pause. OWS became a slogan and nothing more. The Internet animali quietly turned their attention to other mobs.

Why do so many events in the 21st century unfold in such a similar manner? What is the nature of flashes? What is it about society that makes us act like mindless waves on a beach? Do flashmobs have structure? To find out, I built the following simulator, cleverly called, *Flashmob*. Flashmob simulates the behavior of people that are easily influenced by what they read on the Internet, see on TV and YouTube, and hear from friends and neighbors. Their motivations may come from someone else, but are they really under the control of flash-herders like Lasn and his Adbuster tribe? Or do *Internet Animali* have a mind of their own?

2.5 Internet Animali

Why do people boulevard? Why do they rave, flash, and incite? One way to answer this and related questions, is to simulate the very fabric of *social networks*—the invisible forces that hold people together or drive them apart. **FlashMob** is software that mimics the behavior of a rather mindless social network. It models the influence that members of the network have on one another. It is a simplification of reality, but a very interesting model nonetheless, because it shows that under certain conditions, flashmobs are predisposed to act like mindless boulevardiers. Mobs of long-tailed size emerge from randomness with little provocation and a lot of peer pressure. *Spontaneous order* emerges out of chaos.

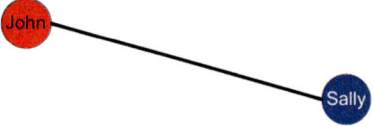

Fig. 2.6 Actors John and Sally influence one another so they are connected by a social link. John currently sides with the *RED*s while Sally agrees with the *BLUE*s

Here is how **FlashMob** works. Each person in the simulation is represented inside of the computer program by a node or *actor*. Influence relationships between pairs of actors are represented by connections called *links*. A link between two actors represents a bi-directional influence. For example, a link between John and Sally represents John's influence on Sally and Sally's influence on John. See Fig. 2.6. Directly connected actors are called *neighbors* so only his or her neighbors influence an actor.

Now suppose each actor takes one of three possible *positions* on a topic or subject—RED, BLUE, or WHITE. If an actor agrees with the RED group, it is painted red; if he or she agrees with the BLUE group, he or she is painted BLUE; and if he or she has no opinion or position, the actor is painted WHITE. A position can be an opinion, belief, political affiliation, product endorsement, belief of a member of a sub-culture, etc.

Initially, one actor is BLUE, one is RED, and all others are WHITE (neutral). The initial actor's positions remain fixed throughout the simulation. We say they are *pinned*. Regardless of the opinion of their neighbors, pinned actors hold onto their position—either RED or BLUE. We are interested in the effect these stubborn actors have on the rest of the network. Can they persuade others to adopt their positions? Can one actor start a revolution?

The goal of the simulation is to determine the eventual outcome from spreading the pinned actor's influence to all other actors in the network through interactions with their neighbors. For example, friends on Facebook.com or tweets on Twitter.com exert an influence on each other through postings. Similarly, the opinions of consumers on Yelp.com influence the popularity and desirability of products and services. Do these influences spread to everyone on the network?

Neighbors influence actors but actors also have their own level of *conviction*—a number between zero and one—that determines how difficult it is to change their minds. A conviction of 1.0 means the actor never changes his or her position; zero means he or she is completely influenced by neighbors; and 1/2 means the actor sides with neighbors one-half of the time and ignores them the other half. Stronger conviction means it is more difficult to change actor positions through influence.

One actor is painted RED and another actor is painted BLUE, initially. Then the simulation performs a very simple rule over and over again until it is stopped. It paints each actor RED with probability $(1 - C)R/(R + B)$, where C is conviction, R is the number of links connecting the actor to a RED actor, and B is the number of links connecting the actor to a BLUE actor. For example, if conviction is zero,

2.5 Internet Animali

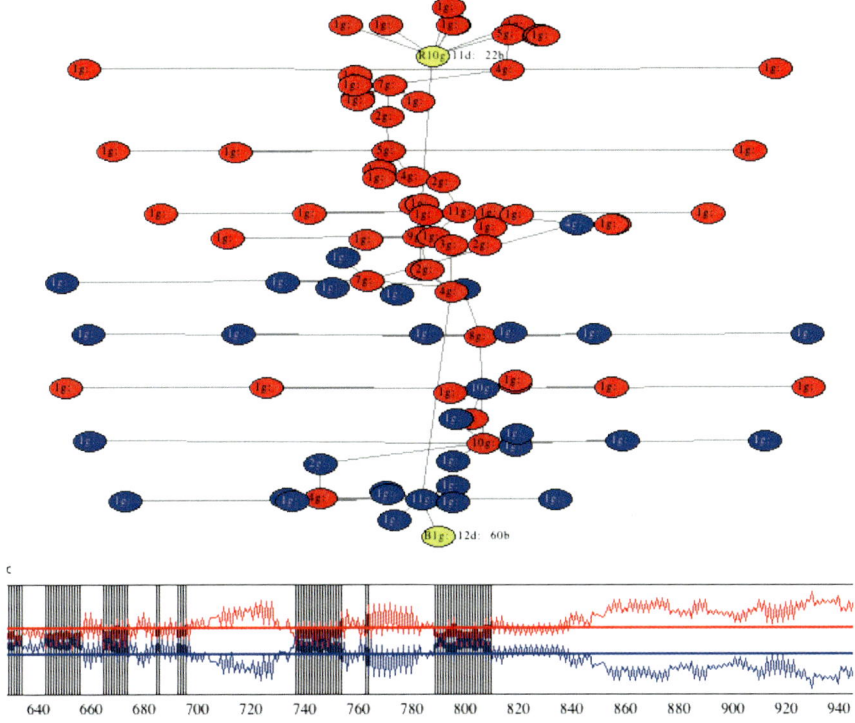

Fig. 2.7 A social network forms groups—mobs—around *RED* and *BLUE* positions. The strip chart at the *bottom* of the simulation display shows the change in mob size versus elapsed time. *Vertical lines* mark points where the sizes are equal or cross each other

and an actor has 3 RED neighbors and 4 BLUE neighbors, then the actor is painted RED with probability 3/7. Otherwise, the actor is painted BLUE.

Starting with two pinned actors **FlashMob** determines the color of each actor as it steps through time. The number of RED actors oscillates, growing and shrinking somewhat chaotically and without an apparent direction. BLUE actors do the same. Once in awhile the number of RED (or BLUE) actors will surge—pull ahead of the opposition and dominate. **FlashMob** records the elapsed time between RED domination and BLUE domination. RED/BLUE actors dominate the social network when the largest connected sub-network of one color exceeds 50 % of all actors. This domination switches whenever the majority switches from RED to BLUE or BLUE to RED.

RED and BLUE contagions spread through the network like the plague, modifying actor positions according to their convictions and neighbor's positions. Figure 2.7 illustrates a stage in the propagation of RED and BLUE positions as the simulation steps through time. Figure 2.8 shows the distribution of size-of-largest-connected sub-network of RED and BLUE groups for a typical simulation run.

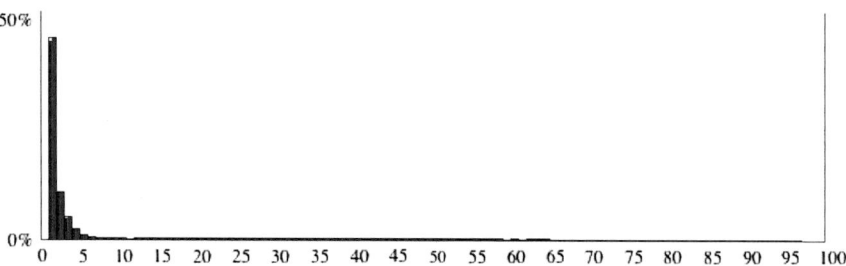

Fig. 2.8 Simulating the formation of flashmob spontaneously formed by listening to neighbors leads to a long-tailed size-distribution

Clearly, the long-tailed distribution of Fig. 2.8 is quite similar to the size-distribution of real flashmobs as shown in Fig. 2.1.

Flashmob illustrates a fundamental truth: social networks form spontaneously and self-organize without any help from outside. Order emerges out of disorder and then dissipates almost as quickly. This emergence is bottom-up, not top-down. That is, the order comes from local actions, rather than from a designated leader or group organizer. **Flashmob** organization is driven by conviction—or lack of it. Left to its own devices, social networks form and dissipate, automatically.

What else does **FlashMob** tell us? First, assuming actors lack conviction and are easily swayed by neighbors, the size of the largest group of RED or BLUE actors obeys a long-tailed distribution just like the real flashmobs analyzed in Fig. 2.1. The elapsed time between the formation of a RED and BLUE majority, or the reverse, obeys a Levy Flight very similar to the Levy Flight distribution shown in Fig. 2.3. At least with respect to size and elapsed time between RED/BLUE majorities, **FlashMob** replicates the behavior of real flashmobs observed in the wild. In other words, there seems to be no difference between the behavior of a mindless computer simulation and real humans forming real social networks around real issues. But, this result depends very much on where the first RED and BLUE actors are placed in the network.

Suppose the same social network is seeded with different initial RED and BLUE actors. In this scenario, the pinned RED actor is connected to twice as many neighbors as the pinned BLUE actor. In a real sense, the RED actor has more influence over the entire network simply because he or she is linked to more neighbors. This small advantage is critical, because it magnifies as RED's influence spreads through the entire network.

In this lop-sided case, the long-tailed size distribution dissolves because many more RED mobs form than BLUE mobs. RED becomes dominant. The domination of RED over BLUE becomes even more pronounced as conviction rises. In fact, the long-tailed distribution of mob size completely disappears and is replaced by two Normal distributions—one centered on the average RED mob size, and the other centered on the average BLUE mob size, see Fig. 2.9. Domination of one position over another dissipates flashmobs.

2.5 Internet Animali

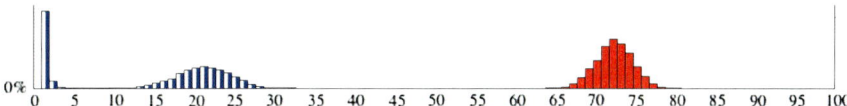

Fig. 2.9 Distribution of *RED* and *BLUE* mob sizes when the initial *RED* is connected to 8 neighbors and *BLUE* is connected to 4 neighbors, and conviction is 1/2

The previous simulations were run with 3–4 links per actor, so they modeled a *sparse network*. [The average number of friends on *Facebook* is typically 130–150]. Now suppose the social network is *dense*, meaning that many neighbors influence a typical actor. In this scenario, the average number of links per node is 20. How do mobs form in a dense network? As it turns out, flashes disappear once again. The sizes of both RED and BLUE mobs obey Normal distributions centered on 50 % of the actors. Flashmob size is purely random and no pattern emerges. Flashing is limited when actors listen to too many neighbors. Nonetheless, the elapsed time between RED and BLUE majorities still obeys a long-tailed distribution.

2.6 Mob Power: The Hub

These simulated social networks lack one more property observed in real social networks: a *hub*. In almost all social networks there is one actor that has far more connections than the average. This highly connected actor is called a *hub*, for obvious reasons. The social network is no longer randomly connected. Instead, it has structure in the form of a few highly connected actors and many less connected actors. Lasn and White were hubs in the *Adbuster* network, because they were "connected" to 70,000 subscribers. Their influence far outweighed all other actors. What impact does a hub actor have on the rest?

Assuming no conviction and the number of neighbors of the pinned RED actor is twice that of the pinned BLUE actor, the size distribution of mobs is still long-tailed as before. But the tail of RED is much longer than BLUE. In fact, the size of the largest BLUE mob rarely exceeds 10 % of the population. The initial position of the pinned hub dominates the positions of nearly all actors. The effect on RED is to elongate its long-tailed distribution. The effect on BLUE is to radically shorten its tail. In other words, network structure introduces extreme groupings. Pinning a hub introduces polarization of the network. The hub exercises social control over the mob.

Power lies in self-organization of social networks. As a network formed by RED actors organizes, it gains more influence over the BLUE actors. The RED network becomes dominant whenever it can self-organize faster than the BLUE network. Self-organization can be measured in terms of a network property called the *spectral radius*. Therefore, power over the entire network rises with an increase in spectral radius. If the spectral radius of the RED network exceeds the spectral radius of the BLUE network, RED will dominate BLUE. In other words, the rate of self-organization determines which position wins.

What do these simulations tell us about Internet animali? I summarize the results shown in Fig. 2.7 through 2.9:

1. Flashmob formation is a Levy Flight in size and elapsed time between switches in majority opinion when the social network is *sparse and conviction is low*. Low conviction accelerates the rate of self-organization. Large flashmobs quickly form when actors lacking conviction listen to a small number of strong influencers. On the contrary, large flashmobs are inhibited from forming when conviction is high or actors listen to more than a dozen neighbors. Too many influences attenuate flashmob intensity and retard the rate of self-organization.
2. A pinned RED actor dominates BLUE actors if its number of neighbors exceeds the number of neighbors of BLUE, and vice versa. If you want to steer mobs in one direction, start with the most connected actor. Revolutions are started and successfully completed by highly connected and strong actors simply because number of links equals influence. A small advantage in terms of connectivity accelerates the rate of self-organization.
3. Structured networks with a large pinned hub are dominated by the initial position of the hub. This structure introduces additional extremes—the tail of the hub distribution is longer, and the tail of the subordinate actor distribution is shorter. The larger the hub, the easier it is to control the network and the larger is the flashmob. A hub amplifies influence—a scale-free network amplifies influence of the hub even more.[4] Self-organization—as measured by spectral radius—is enhanced by hub size.
4. Structured networks with large hubs change majorities more often when the pinned initial RED and BLUE actors have more links than when they have fewer links. Hub size introduces volatility. Conversely, fewer links means longer elapsed times between majority switching. Pinning highly connected actors introduces more volatility, and pinning slightly connected actors introduces stronger majorities. When conviction is low, flashmobs can be whip-lashed back and forth by highly connected actors. Self-organization goes hand-in-hand with hub size, setting up competition between RED and BLUE when both networks contain large hubs.

Clearly these simulation results have important applications. Whether in advertising, marketing of web sites, promotion of political campaigns, or simply understanding the dynamics of socio-political movements around the globe, flashmob behavior modification is a matter of pulling the right levers. Mob formation has little to do with human nature and psychology, and much to do with conviction and the inclination of your neighbors. Power and influence is a mechanical property of boulevardiers, activists, and trendsetters. Persuasion can be had simply by shaping the topology of the network. Function follows form.

---

[4] Recall: *a scale-free network* is one with a few large hubs and many actors with few connections.

2.7 Put Out the Fire

On May 25, 2011, Mongolian students marched in protest before a government building in Xilinhot, Inner Mongolia. Reports on the Internet of the fatal beating of a street vendor triggered a flashmob of hundreds of people in the southern China city of Guangzhou. In September 2011, riot police removed protesters from the entrance of a Solar panel manufacturer in Haining, Zhejiang province. The factory is a subsidiary of a New York-listed firm.

Thousands of outraged residents of Wukan, an urban village located in Guangdong province, held daily protests over the death of a local man held by police. The villagers were protesting the seizure of their land to make way for industrial development. They have complained of government land grabs for decades, but a massive real estate project announced in September led to an outpouring of pent up anger resulting in riots and clashes with police.

The number of Chinese "mass incidents" rose from 8,700 in 1993 to 74,000 in 2004, more than 90,000 in 2006, and then to 180,000 by 2010, according to the Chinese Police Academy and Sun Liping, a Tsinghua University sociologist.[5] The proliferation of cell phones and social networks like Facebook and Twitter has exponentially increased the number reported. The minute a window is smashed, somebody shoots a video and posts it on a blog.

These flashes were sparked by claims of unpaid wages, taxes, lay offs, land seizures, factory closings, poor working conditions, environmental damages, corruption, misuse of funds, ethnic tensions, use of natural resources, forced immigration and police abuse. Some of them have been quite large and violent. However, most were too small to gain widespread notice. I haven't done the math, but my guess is that these are long-tailed events.

China has long been associated with repression of its population, censorship, and dictatorial rule. So why the recent change? Why does China tolerate even the smallest uprising? Have the Chinese rulers gone soft?

2.8 Forest Fires

In my previous book, *Bak's Sand Pile*, I showed how large forest fires are prevented by frequently setting small fires or purposely removing trees. When swatches of forest are removed from the stockpile of fuel, they form natural barriers to the uncontrolled spread of fire. In fact, modern day forest managers practice orderly deforestation as a means of reducing the size of wild fires. This is called *depercolation* in the scientific literature—tree growth is *percolation*, and tree removal is *depercolation*. Depercolation reduces self-organized criticality so that forest fires are less deadly.

---

[5] http://chinadigitaltimes.net/china/sun-liping/

Does the same technique work for the Chinese government? By allowing a lot of relatively small protests (small fires), does China avoid the really big ones that could bring down the government (big fires)? Is there such a thing as "political depercolation"? This is left as an exercise for the reader.

What we know for sure is that if you want to stop a flashmob, you have to attack its hubs—the most highly linked actors. Flashmobs are long-tailed phenomenon and the larger the hub, the longer the tail. Quelling the activities of hubs shortens the tail. Completely eliminating the hub renders it almost insignificant. Similarly, social networks containing actors with a shared conviction also shortens the tail. It is very difficult to ignite flashes in a population of people with strong convictions. This is the key to governance in the 21st century where governments must walk a tightrope between anarchy and mob rule in the age of the global Internet.

But flashmob psychology and social network effects are just one tip of a much larger iceberg. To see the ice under the surface, we need to understand what ignites a flashmob. This requires an understanding of Sparks.

References

Lewis, Ted G., Bak's Sand Pile, AgilePress, 2011. 378 pp. http://www.agilepress.com/index.html
Schwartz, Mattathias, Pre-Occupied, The New Yorker, Nov. 28, 2011, pp. 28–35

Sparks

Abstract

What sparks a flashmob? What starts the emergence of a self-organized social movement, political change, technological breakthrough, and other chain-reactions leading to the rise or collapse of a new milieu? Ideas, movements, and products are sparked by a highly evolved sense-making mechanism that has emerged over the past decade to make sense of big data—copious volumes of online documents, email, tweets, and probes collected via the Internet. This data has to be reduced by analyzing its wave-like and probabilistic behavior. In a highly connected interdependent world, ordinary probability is replaced by conditional or Bayesian probability, which in turn leads to Levy flights and long-tailed events widely ranging from making sense of big Internet data to finding and killing Osama bin Laden. Thus, the 21st century is an era of massive data processing unheard of just a decade ago. This transformation leads to an unanswered question, "Is the human race on the precipice of making an intellectual leap of profound significance because of this massive collection effort?"

3.1 A Blip in the Night

Sohaib Athar, a 33-year-old Senior Partner at Algotrek Technology Consulting in Abbottabad often retreats into the mountains with his laptop computer and Twitter account to get away from life's cacophony in Pakistan. Early one Sunday morning in May 2011, while enjoying the peace and quite of the night, he heard a "window-shaking boom" a few miles away.[1] He tweeted, using his Twitter alias *Really-Virtual*, "I hope it's not the start of something nasty." But the fireworks got worse

---

[1] https://twitter.com/ReallyVirtual

as four stealth helicopters swept into a terrorist compound in the valley below. *ReallyVirtual* tweeted, "Go away helicopter—before I take out my giant swatter".[2] The bump in the night was one of the four helicopters crashing moments before the assassination of public enemy number one—Osama bin Laden.

Athar was the first "reporter" to witness the raid on Osama bin Laden's compound, and the first to tell the rest of the world. Later, Athar tweeted, "Uh oh, now I'm the guy who *liveblogged* the Osama raid without knowing it." Soon he had 61,829 Twitter followers—and even worse - fame. More to the point—Athar published the top-secret raid 10 h before anyone in the media even knew what was going on. His tweet may have been a blip in the night, but it also said something important about the 21st century—not even top-secret raids are immune to the ever-watchful eyes of the Internet. Forget about privacy.

Athar's prowess illustrates a key ingredient in our highly interconnected universe—the Internet, and the people that populate cyberspace—form a highly instrumented sensor network that spans all of humanity. Over a billion people with laptops, cell phones, and tablet computers now feed raw data into a realtime, omnipresent, omniscient *firehose* so big and wild that no single company, news media conglomerate, government, or United Nation can fathom it, let alone control it. The accessibility of everyman/everywoman *big data*, as it is called, is unprecedented in human history. It portends to be revolutionary and disruptive.

But *data* is not the same thing as *information*. The problem with omnipresent, omniscient data is that it is big, noisy, and often hides rather than exposes, the blips in the night. Copious volumes of data may not make sense. And, knowing everything about everyone may not always turn out to be the best thing for freedom-loving people. Capturing data may mean the end of privacy and security as we know it. Big data may be a double-edged sword.

3.2 Knowledge Is Power, but Knowledge Is not Big Data

In his blog/book, *The Science of Social*, Michael Wu defines *information* as what is left over after raw data is compressed.[3] For example, my previous book and all of its supporting references, figures, and revisions takes 555 MB of disk space in its uncompressed human readable form. After compressing it into a *zip* file, it consumes 140 MB—almost one-fourth as many bits. Wu's metric says the 555 MB of raw data in the human-readable version of *Bak's Sand Pile* actually contains only 140 MB of information. What is the remaining 415 MB for?

Compressing a raw data file removes redundancy and meaningless noise. Therefore, information is data *minus* redundancy and noise. It is just the facts, stripped of embellishments, duplication, and nonsense. It is the content of this

---

[2] Bell (2011).

[3] https://lithosphere.lithium.com/t5/Science-of-Social-blog/bg-p/scienceofsocial

chapter without the stories and anecdotes. Information is barebones meaning stripped of all excess. It is unembellished fact.

The challenge of big data is to remove the meaningless bits by filtering, analyzing, and interpreting large volumes of realtime data as it is created by the media, intelligence agencies, corporate IT departments, etc. In the case of the Osama bin Laden raid, the raw data that the US military had to sift through was like finding the proverbial needle in the haystack of telephone calls, rumors, sightings, and informer interviews. Compared to bin Laden's geographical location, everything else was meaningless noise that had to be filtered out and discarded. But which bits are information and which are noise?

In 2011 *Google.com* and *YouTube.com* used 2.25 GW-hrs of electricity to power its servers.[4] These servers processed 100 billion *pageviews* per month. Facebook.com registered one trillion pageviews in the month of June 2011. Google.com, YouTube.com, Facebook.com, Twitter.com, and iTunes.com accounted for more than 95 % of global Internet traffic. Their users are the new heartbeat of humanity—the human sensors that generate big data in the form of pageviews, blogs, tweets, posts, uploads, downloads, and commentary. But, most of it is useless noise!

How big is big data? In 2010, the Internet contained an estimated 10^{21} *bytes* of raw data, which is equivalent to all of the facts, figures, images, emotions, and memories of 20 million human brains.[5] And assuming data collection doubles every year, the Internet should contain the equivalent of all human brains on the planet by 2022—all 7.5 billion of us. If we compress and remove the redundancy, the sum total of all information possessed by humanity will equal about 10^{24} bytes (yottabytes) of raw data which when compressed, still exceeds 10^{23} bytes (kilo-zettabytes) of information. These numbers are so big the human mind cannot comprehend them.

Assuming each of these human brains tweets or googles once per day, the realtime meaningful information flow equals 2 billion "facts" per day, or 23,148 facts per second, every second of every day. But of course, these 23,148 facts must be plucked from the ether of cyberspace—from the kilo-zettabytes of data pouring in from all over the globe. The Internet firehose spews out so much raw data that only a building full of computers can process it in time.[6] As it turns out, application of unprecedented computer processing power is exactly how governments and corporations are exploiting the Internet firehose.

---

[4] 2.25 GW-hr is enough to power a city with 200,000 homes.

[5] John Gantz, IDC. www.emc.com

[6] In 2012, Twitter's firehose, alone, hosted 340 million tweets per day.

3.3 Wizard of OS

Billions of human sensors constantly dump raw data into the Internet every second of every day. But to make sense out of it, these raw tweets, searches, commentary, downloads, and uploads must be processed to squeeze out meaningful information. No single computer is powerful enough to do this, so Doug Cutting built *Hadoop*—an open source file system named after his son's toy elephant. Hadoop evolved from Google's MapReduce specification, which described how to break down massive amounts of raw data into pieces, distributing the pieces to thousands of computers, and then processing the pieces in parallel. For example, in 2009, Yahoo.com announced indexing 21 petabytes of data running MapReduce with Hadoop on 10,000 computers.[7]

Doug Cutting took time off from his Silicon Valley day job to write Hadoop, and now it is an industry-wide infrastructure, supported and used by nearly every major Internet corporation in the world—Amazon, American Airlines, Apple, EBay, Federal Reserve Board, IBM, Hewlett-Packard, LinkedIn.com, Microsoft, NetFlix.com, Yahoo, and the New York Times. Hadoop provides the infrastructure for 21st century *sense-making*.[8] Some technologists equate Hadoop with cloud computing—the current focus of large corporations, governments, and spy agencies.

The New York Times, for example, reported using 100 Amazon cloud computers to "understand" and convert 4 terabytes of raw image data into 11 million PDF files at a cost of $240.[9] But it took 24 h to do the job. Four terabytes is relatively small data. Nonetheless, it is a beginning. Even with the processing power of thousands of computers harnessed together to convert raw data into meaningful information using MapReduce and Hadoop, the challenge of big data is extreme. It requires amassing huge hardware and software assets.

Apache Hadoop 1.0 is the open-source "completed project" that was released in 2011. In March 2011, Apache Hadoop won the Media Guardian Innovation Award, called the Megas, beating out the *iPad* from Apple Company. Hadoop was called the "Swiss army knife of the 21st century" by the judges, who exuded, "[Hadoop] has the potential to change the face of media innovations." Cutting created an industry as a byproduct of spending more time with his son, and like Sohaib Athar, accidentally stumbled onto fame.

3.4 Complex Crowd Sense-Making

Governments and corporations are working hard to turn *crowdsensing* into actionable information. Crowdsensing is the process of collecting raw data from disparate sources and "cleaning" it prior to making sense out of it. Typically it

---

[7] 21 petabytes is 2.1×10^{16} bytes.

[8] Sense-making is the process of converting the experiences of many people into meaning.

[9] http://en.wikipedia.org/wiki/Apache_Hadoop

means gathering large amounts of raw data from sources like Facebook and Twitter, removing the noise, and converting bits into meaning. It sounds simple, but it isn't.

MapReduce and Hadoop can only turn raw data into actionable information if they are combined with *smart processing algorithms*. These algorithms are generically called *analytics, predictive analytics*, and various other names. They are a basket of computer algorithms designed to find patterns, classify structured and unstructured (text) data, thus converting it into information. [Information may still lack meaning].

Here are a few examples of crowdsensing and corresponding analytics:

Turn Tweets (data) into favorable product sentiment (information).
Predict power consumption (information) from 350 billion annual meter readings (data).
Discover one credit card fraud (information) from among 5 million transactions (data).
Reduce consumer churn (information) by analyzing 500 million daily telephone calls (data).
Monitor live video feeds (data) to prevent terrorism (information).

According to an article in the online *New York Times*, "When *Wal-Mart* wanted to know whether to stock lollipop-shaped cake [baking gadgets] in its stores, it studied Twitter chatter. Chatter big data is converted into smart inventory management. *Estée Lauder's* MAC Cosmetics brand asked social media users to vote on which discontinued shades to bring back into stores. The stuffed-animal brand *Squishable* solicited Facebook feedback before settling on the final version of a new toy. And Samuel Adams asked users to vote on yeast, hops, color and other qualities to create a beer, an American red ale called *B'Austin Ale*. The consumer-created brew got rave reviews".[10]

Anyone with an Internet connection can perform crowdsensing research through online services like Topsy.com.[11] Topsy takes Google-like search terms as input and returns a plot of number of mentions, amount of influence, and degree of sentiment versus the dates when readings were taken from the firehose. For example, entering "iPad" and "Kindle" into the Topsy home page produces a line chart showing that "iPad" was mentioned twice as often as "Kindle" during the month of September 2012. "iPad" searches are further broken down by influence source (webpage), and Topsy-defined metrics such as influence, momentum, velocity, and peak.[12] Topsy.com does the crowdsensing and analysis for you.

---

[10] Clifford (2012).

[11] http://analytics.topsy.com/ was purchased by Apple Inc.

[12] Influence, momentum, and peak are relative terms defined by Topsy. Influence relates the power to influence purchasing decisions; momentum is roughly a measure of inertia (rising or falling); peak is the point of maximum influence.

3.5 21st Century Sense-Making

This level of simple crowdsensing is a holdover from the 20th century. It can identify major trends and shifts in sentiment and influence, but it does not pinpoint the blips in the night, nor explain what they mean. That is, simple crowdsensing summarizes massive amounts of data, but tells us almost nothing about emergent ideas, products, or services before they become mainstream. It can record and calculate averages, but it cannot explain and predict outliers. It provides sensing, but not sense-making. Crowdsensing cannot predict the future.

The real power of big data is in its ability to "read between the lines" and explain the previously unexplainable. In the 21st century, big data will increasingly be used to deduce what is *not* explicitly present in the data. This is a step above crowdsensing, because it yields insight. I'll call this *complex crowd sense-making*. Complex crowd sense-making turns information into knowledge—revealing not only facts and figures, but also how pieces of information fit together like a puzzle. What facts stimulate emergent behaviors like flashmobs? How might analysis add value as information is accumulated and combined? Complex crowd sense-making borders on *epistemology*.[13]

Take the spread of malaria as an illustration of how to glean insight regarding the spread of a deadly disease by analyzing seemingly unrelated raw data, mashing it together, and adding analysis.[14] Malaria kills over 650,000 people every year, so eradication of malaria is a serious problem. But how can big data cure a disease?

A study led by researchers at Harvard School of Public Health and seven other institutions correlated the movement of malaria-infected people with GPS traces from cell phone calls. They found that malaria emanates from the mosquito infected Lake Victoria region and spreads east to Nairobi, where the high concentration of people accelerates the spread of the contagion. Nairobi is a kind of force-multiplier hub. After contracting the disease, the cell-phone toting carriers return to their villages where they infect others. This process repeats, accelerating the spread with each cycle.

What does cellular telephony have to do with the spread of infectious disease? Quite a lot, as it turns out. The researchers mapped every call or text made by 14,816,521 Kenyan mobile phone subscribers to one of 11,920 cell towers in 692 different settlements. They were able to identify the flow of malaria germs through human sensors carrying cell phones. The telephone companies collected big data, and the researchers added analytics to create new insights. Raw data was turned into information. Information was turned into meaning. Meaning was turned into actionable policy to fight malaria.

---

[13] Epistemology studies the nature and scope of knowledge.
[14] Wesolowski et al. (2012).

3.6 Astroturfing

Every positive use of technology also has its negative uses. *Astroturfing* is the practice of spreading misinformation through social media like Twitter and Facebook using automated software called a *bot*. A bot is a kind of virus that spreads through the Internet infecting servers and home computers alike. When used to misinform people, bots become a negative technology.

For example, political elections are a prime target of astroturfing because specialized bots can convince people to vote (or not vote) for a certain candidate based on fraudulent "popular consensus". Online "human consensus" can be manufactured by a cleverly programmed bot that periodically spews out fake tweets.

The reason bots are called bots—short for robots—is that a bot can be programmed to simulate the behavior of a human tweeter. It can sleep when humans typically sleep and manufacture thousands of "human-sounding" tweets like, "candidate X cheated on her income tax returns", or "country Y violates human rights". A Twitter bot can take on the identity of hundreds of hash codes to simulate the tweets of a large number of humans. A big data analyzer may mistakenly register these artificial tweets as authentic human sentiment.

Filippo Menczer, Professor of Informatics and Computer Science and the Director of the Center for Complex Networks and Systems Research at the Indiana University School of Informatics and Computing, Bloomington, claims to be able to convert caffeine into science and is responsible for the term "astroturfing".

His center stood up a web site that tracks politically motivated tweets called *Truthy*.[15] He says astroturfing can come off to the untrained eye as a vitriolic surge against a candidate by the grassroots community and quickly evolve into a real concern by a candidate's constituency. But it is machine-made artificial grass. Can a piece of software spin a yarn that changes voter's minds at the polls?[16] Menczer says yes, and leads a group of researchers to better understand the spread of misinformation through social media.

Astroturfing points to the need for more sophisticated techniques that avoid misinformation and find truth wherever it may be hiding. Big data analytics must dig deeper. Reasoning about the Internet world requires more than simply counting tweets. Perhaps "reasoning algorithms" based on artificial intelligence technology are the answer. According to some, this calls for an old technology that was lost for nearly 300 years, and only resurfaced in the past decade or so.

Sometimes what used to be old becomes new again.

---

[15] http://truthy.indiana.edu/
[16] Menczer (2012).

3.7 An Old Technology

One of the most powerful analytic tools of the 21st century comes to us from an 18th century holy man, Thomas Bayes (1701–1761). Bayes was a Presbyterian minister and amateur mathematician. Like Doug Cutting, Bayes escaped from his day job as a preacher of dissident theology to think deep thoughts about the uncertainty of his emerging modern world. After all, the Industrial Revolution was in its formative stages by 1760, much like the nascent Internet is in its formative stages today. Such ferment makes people think deep thoughts. Bayes came from a wealthy Sheffield, England family who enjoyed the success of the famous Sheffield cutlery industry—an early Industrial Revolution business. He had the time and resources to be an outlier.

Unfortunately, Bayes' ideas were 300 years too early. His most valuable work was published after his death, and remained in the backwaters of science for the next three hundred years.[17] A close friend, Richard Price, collected and edited Bayes' notes and published his most important contribution—the famous and controversial *Bayes' Theorem* on "inverse probability".

Bayes' Theorem is simple, but has deep philosophical implications. In earlier chapters I refer to the extreme connectivity of the 21st century and claim that the likelihood of earth-shaking events are conditional—meaning that the next surprise to hit society is dependent and conditional on previous events that may have occurred on the other side of the globe. Pre-cursor events increase the likelihood of subsequent events, according to Bayes' Theorem:

$$\Pr(A \mid B) = \frac{\Pr(B \mid A) \Pr(A)}{\Pr(B \mid A) \Pr(A) + \Pr(\textit{false\_positive})}$$

$$\Pr(\textit{false\_positive}) = \Pr(B \mid \textit{notA}) \Pr(\textit{notA})$$

The left-hand side, Pr(A | B) is the *posteriori probability* of event A, conditional on event B.[18] The right-hand side terms, Pr(B | A), Pr(B), and their complements Pr(B | notA) and Pr(notA) are *priori probabilities* based on "beliefs". Actually, a priori probability is a *degree of belief*—a measure of how confident we are that a certain fact is actually true. Therefore, Pr(A | B) is conditional on (historical) evidence of Pr(B | A) and Pr(A), and complements Pr(B | notA) and Pr(notA). The product of complements Pr(B | notA)Pr(notA) is the likelihood of a *false positive*. That is, in any belief system there is a certain probability that we are wrong. When false positives are reduced or removed, certainty is increased and so is the degree of belief in the Pr(A | B). Bayesian probability accounts for erroneous beliefs such as a false positive.

---

[17] Bellhouse (2004).

[18] Posteriori means "after" and priori means "before".

3.7 An Old Technology   43

If it is known that B has occurred in the past because of A, this knowledge increases the belief that A will happened in the future. Furthermore, if false positives are exposed for the fraud that they are, this knowledge increases the belief in A even more. As evidence piles up, uncertainty is reduced, and confidence in likelihood Pr(A | B) also increases.

Bayesian logic is so confusing that an example is needed to understand the power of Bayes' simple, yet deep, theorem. Bayesian logic becomes even more confusing when complex relationships among factors—called *propositions*—are incorporated into a model and then connected to big data.[19] To get the most out of Bayes' Theorem, we need a way to connect propositions together and mash them up with big data. How do these pieces fit together?

3.8 Bayes' Network

Suppose a company wants to measure the effectiveness of online advertising. Further suppose the company has access to big data, and the computing resources to analyze massive amounts of big data in real time. What kind of analytics should the company use? One of the favorite analytic tools of Internet companies is a *Bayesian network*—a model of cause-and-effect based on Bayes' Theorem. I will abbreviate Bayesian network as BN, and use a simple BN to illustrate how big data and analytics might be used to elicit insight into the effectiveness of an advertising campaign on sales of the hypothetical Internet company.

A BN is a set of propositions (*nodes*) and influences (*links*) that model crowd sense-making—or any other perceived reality. It is a hypothesis to be tested by analyzing, and concluding causes and their effects observed in actual data. Figure 3.1 illustrates the idea. Each proposition uses Bayes' Theorem to transform historical data into degrees of belief expressed as probabilities.

Propositions, and therefore Bayes' equations containing conditional probabilities, are linked together in a network of relationships. A proposition is a parent if a link connects it to a child proposition. Beliefs flow from proposition to proposition through links, according to Bayes' Theorem. [The child's beliefs are influenced by the parent's beliefs]. Whenever a conditional probability changes, it affects other probabilities through the linked relationships among propositions.

The BN of Fig. 3.1 hypothesizes that product features, peer pressure from friends, and popup advertising influences the buying decisions of online shoppers. Shoppers either buy because of these three influences or wait for more evidence. This binary choice, and its likelihood of happening, is shown in the Decision proposition box of Fig. 3.1. Similarly, the likelihood of Liking or Disliking a certain product is assigned an initial probability in the Features and Friends propositions. And the likelihood of showing or not showing a popup ad is initially assigned to the Popup proposition box.

---

[19] Proposition: A "fact" proposed for acceptance as "truth".

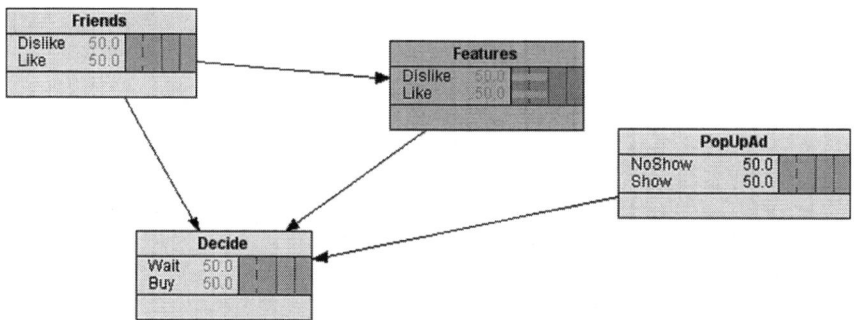

Fig. 3.1 A simple BN model of how buyers are influenced via the Internet. This model hypothesizes that a product's features, comments from friends, and popup advertising persuade buyers to purchase the company's product, according to degrees of belief rendered as conditional probabilities

Initially, the company believes there is an equal chance (50–50 %) of being influenced by any of the three input propositions, and therefore, an equal chance of buying. This is based on zero information and should be considered unreliable. But if big data is mined for more information, the initial belief system can be improved. This is where big data is mashed up with the model.

Suppose the company uses a variety of crowdsensing techniques to find out what people are doing online. The company might use Twitter, Facebook, Google, and other sources on the Internet, and MapReduce/Hadoop, to convert these firehose feeds into probabilities that the company can use to get better estimates of the degree of beliefs needed to drive the company's BN model.

For example, suppose the big data MapReduce analysis concludes the following, summarized here in tabular format. [Only the "conditional probability table" for the Decision proposition is shown here. There are three more tables—one each for Friends, Features, and Popups. Every proposition contains a conditional probability table with a row for every possible combination of inputs].

Friends	Features	Popups	Wait	Buy
Dislike	Dislike	NoShow	50	50
Dislike	Dislike	Show	45	55
Dislike	Like	NoShow	25	75
Dislike	Like	Show	20	80
Like	Dislike	NoShow	75	80
Like	Dislike	Show	80	20
Like	Like	NoShow	5	95
Like	Like	Show	0	100

3.8 Bayes' Network

This table considers all possible combinations of influence on a buyer from the three parent propositions, and their likely outcomes—Like/Dislike, Show/NoShow, and Wait/Buy. For example, if Friends Dislike the product, but the consumer Likes the product's features, the decision to buy rises from 50 % to 75 % without a popup ad, and 80 % with the popup ad. Alternatively, if Friends Like the product, the chances of buying increase further to 95 % without an ad and 100 % with an ad.

The BN analyzes these beliefs by applying Bayes' Theorem to the entire network. The math is messy, but the outcome is insightful. If the popup ad is randomly shown 50 % of the time, the likelihood of buying the product is 72 %. But if the ad is shown (and seen) every time, the likelihood of buying increases to 78 %. Therefore, the popup ad campaign increases sales by 6 %. Assuming 100,000 purchases per month before the ad campaign, the company should expect 106,000 purchases with the popup ads.

But this is not the end of the story, because consumer sentiment is constantly changing in real time. Therefore, the assumptions underlying the BN change in real time, too. What if the impact of Features and Friends on buying decisions also changes? For example, if features suddenly become more influential than before, the likelihood of buying can be quantitatively evaluated by running the BN model again and determining the impact of a feature on sales. The degrees of belief are changed in the BN model and the analysis repeated.

Bayes' Theorem is an exotic and mathematically sophisticated model of reality. It combines evidence-based logic with probability theory in a practical way. As the confidence in the likelihood of pre-cursor events increases because of accumulated evidence, uncertainty in the likelihood of a future event declines, yielding more "belief" in Pr(A | B).

It is no surprise that BN models are used in everything from predicting terrorist attacks, diagnosing patients' diseases from symptoms, to steering an automobile. With a little bit of engineering, a BN can be trained to avoid astroturfing and other artificial artifacts of the social media stream of consciousness. Can this kind of artificial intelligence evolve into a sentient Internet? I address this provocative question at the end of this chapter.

3.9 Google's Car

Judea Pearl (1936–) coined the term *Bayesian network* in 1985 and did much of the early work on *probabilistic inference*—a more general term for Bayesian reasoning. The Tel Aviv-born American prodigiously published papers on machine learning throughout the 1980s. In 2011 he received the *A. M. Turing Award* from the *Association of Computing Machinery* for this pioneering work:

> He is credited with the invention of Bayesian networks, a mathematical formalism for defining complex probability models, as well as the principal algorithms used for inference in these models. This work not only revolutionized the field of artificial intelligence but

also became an important tool for many other branches of engineering and the natural sciences. He later created a mathematical framework for causal inference that has had significant impact in the social sciences.[20]

Throughout the 1980s and 1990s researchers in the field of *Artificial Intelligence* slowly evolved machine learning from an application of rigorous logic to an application of probabilistic inference. They essentially gave up on Boolean logic and turned to a softer form of reasoning based in probability theory and belief networks. They focused on quantifying belief instead of force-fitting their theories into the hard-and-fast rules of logic—a perfect match for BN technology. The ability of a BN to reduce uncertainty was, for these computer scientists, a form of soft learning—a less brittle and more malleable approach.

Perhaps the most dramatic demonstration of applied BN technology came in 2005 when a robotic vehicle named *Stanley* won the *DARPA Grand Challenge*— driving itself 150 miles across the Mojave desert. The Stanford University Racing Team led by Sebastian Thrun (1967-) won in 6 h and 54 min. Thrun went on to lead the development of the Google self-driving car, and became a Google vice president and Fellow. In 2012, with encouragement from Google, Nevada and California legalized autonomous vehicles for use on public roads.

The self-driving car illustrates another type of mash-up between big data and analytics. The Google car relies on Google's massive GPS map data and Bayesian reasoning to find its way through traffic while avoiding collisions with other cars and people. "Since launching in 2007, Google Street View had captured 20 petabytes of data in 48 countries. The company uses cars, 'trikes, snowmobiles and people outfitted with custom cameras to capture 360-degree images around the world."[21] Google's self-driving car matches street view images with GPS location data, images from an onboard camera, and onboard radar data to navigate from point A to B on its own. Its BN essentially learns to navigate the surface of the earth.

3.10 The Sentient Network

Bayesian networks and other artificial intelligence machine learning algorithms are constantly scanning the Internet to "learn things", using evidence-based techniques illustrated by my simple Bayesian network example and the Google car. The big data available online can be used to train bots to recognize when someone is going to buy something, a particular person is likely to commit a terrorist attack, or a disease is about to break out in a certain area of the country. It is as if the eyes of the Internet are constantly watching and MapReducing big data into actionable information.

Is the Internet becoming sentient? This question has been asked—an answered—by a number of others, so I need not repeat the arguments here. But we do know there are limits to what can be known. The famous *Halting Problem* of

---

[20] http://amturing.acm.org/award_winners/pearl_2658896.cfm
[21] Gross (2012).

Alan Turing says it is impossible to decide if a computer system is secure or even correct. *Godel's Theorem* says there is always an un-provable theorem in any formal system of computation. And clearly, the hardware and software that makes up the Internet is a formal system subject to both Turing and Godel's Theorems.

The un-decidability of every formal system is easily illustrated by the following paradox:

> Sam says, "Hilary always lies"
> Hilary says, "Sam always tells the truth"

Which one is telling the truth? Which one is lying? Assuming Sam is telling the truth implies that Hilary always lies. If Hilary always lies, then her statement, "Sam always tells the truth" is false. Thus, Sam is lying. But this is a contradiction. So here is a simple formal system where it is impossible to decide the truth.

The Internet is subject to the limitations of the Halting Problem, and yet it and its users exhibit "intelligence", in a number of fascinating ways. Collaborative users have solved difficult problems that no single person has been able to solve using the collective intellect of the Internet, and "group think" has demonstrated insights that previously were unthinkable. Take *gamification* as an example.

3.11 Gamification

Crowd sensing and crowd sense making are passive—they combine technology and sociology to make sense of what is "happening out there", but they allow for very limited forms of feedback. The one-way street of the early Web began to switch to a two-way turnpike around 2005, with the ascent of *Web 2.0*—a catchy phrase for describing the emergence of consumers-as-content. Web 2.0 differs from Web 1.0 principally in terms of user feedback. Web 1.0 was a vending machine; Web 2.0 is an interactive video game.

I sat down with Rodrigo Nieto-Gomez in the Carmel Highlands Inn overlooking the blue Pacific Ocean a few weeks after he contacted me to discuss his research. Nieto-Gomez traveled the US-Mexico border in search of answers to his Ph.D. dissertation. He asked, "How does domestic homeland security strategy affect the North American geopolitical balance?" Honestly, I didn't know the answer.

Soon after our Margarita-inspired meeting, Nieto-Gomez pioneered a new kind of educational experience—*gamification* in the service of online education for homeland security professionals. Gamification is the application of game technology to non-game situations. In its most primitive format, a web site is gamified by engaging customers in contests, puzzles, quizzes, and challenges, and then rewarding them with points, badges, recognition, and prizes. Nieto-Gomez used it to answer his homeland security questions.

Nieto-Gomez designed STANCE—a game for learning how to combat terrorists and drug lords operating across the US-Mexico border. STANCE is a computer-assisted role playing simulator designed specifically to teach strategy to homeland security students at the Naval Postgraduate School, Monterey, CA. Here is how it works: students are immersed into a complex adaptive security ecosystem with competing agendas or goals to achieve. They are challenged to survive in an uncertain environment using their wits and lessons learned from readings and seminar discussions. For example, student players take on the role of representatives of the Whitehouse, Mexican Government, Business Lobby, civil liberties organization, Lamo drug cartel, and Badiraguato drug cartel. One side tries to suppress drug trafficking, while the other side tries to expand on it.

According to Nieto-Gomez, "What I have seen is that cops who are involved in the border security effort get to better understand why policy makers cannot do the things that, for them, are no brainers at the tactical levels. They go beyond their corner of the system and get a better view of the complex adaptive nature of the border policy environment. That is the biggest virtue of STANCE, I think."

3.12 Human Stigmergy

STANCE illustrates an educational application of human *stigmergy*—the process of collaboration whereby the state of the system under investigation guides the role players rather than the other way around. For example, in Zynga's popular online entertainment called *Farmville*, you need land before you can plant a garden, and you need to harvest the vegetables before you can sell them. If pests destroy the vegetables, a different response on the part of the role player is called for. Thus, the next move in the game is determined by the state of a player's farm.

STANCE is the cognitive equivalent of Zynga's Farmville—an online game whereby the actions of users depend on the current state of the border between the US and Mexico. Planning is replaced by reactions in real time. Strategies and actions change dynamically as the object of attention changes. Stigmergic collaboration among actors (users) leaves a trace that stimulates the next action by the same or different actor. Subsequent actions reinforce and build on each other, leading to the *emergence* of buildings, ideas, products, etc. Human stigmergy is a form of *self-organization* as it applies to 21st century reality.

3.13 Internet Intelligence

The use of human stigmergy, and other forms of *crowdsourcing to solve complicated problems* raises the question of an emerging online intelligence.[22] It suggests that the collective is smarter than any individual. *Fold.it* is a simple but

---

[22] Crowdsourcing means outsourcing tasks to a group of anonymous people.

cogent example. Volunteers flock to Fold.it/portal to solve complex protein folding problems in a matter of weeks that would take a computer years or even decades to solve. One person might take the solution part way, followed by others who complete the solution after seeing earlier partial solutions. In other words, subsequent moves depend on the state of the partially completed puzzle. The final solution emerges from the self-organizing reality.

3.14 Going to Extremes

Raymond "Ray" Kurzweil (1948-), American author, inventor and futurist studies *transhumanism*—the combination of organic DNA with machine DNA. Think cyborg—the part human, part machine collective seen on *Star Trek: The Next Generation*. He claims that at some point in human-machine evolution, a technological singularity or tipping point will be reached leading to a super-intelligent race of cyborgs.

On the other side of the ledger is Jaron Zepel Lanier (1960-), an American computer scientist and virtual reality pioneer who left Atari Corporation in 1985 (along with Thomas Zimmerman) to build and sell virtual reality goggles and gloves. Lanier's book, *You Are Not a Gadget* (Knopf, 2010), criticizes crowdsourcing and human stigmergy as misleading humanity into one giant *hive mind*—an undesirable result of "Digital Maoism". According to Lanier the Internet is retarding progress and innovation: it is nothing more than a modern form of "mob rule".

In between these extremes is Kevin Kelly (1952-), founding executive editor of *Wired* magazine, and author of *What Technology Wants* (Viking, 2010). Kelly argues that technology is a third kingdom. Rather than evolving into transhumans, humans and technology are co-evolving as separate species in an ecosystem occupied by plants, animals, and technology. He calls this third kingdom *Technium*. Kelly ignores the possibility that these two kingdoms might someday interbreed and produce a new race similar to Kurzweil's cyborgs or Lanier's mindless consumers.

Which of these extremes comes closest to reality? Which one is even possible?

3.15 The Self-Referential Web

Whether future intelligent life is populated by transhuman cyborgs, hive minded or not, or co-existing plants, animals, and technium, depends on whether any organic or inorganic entity can achieve consciousness on par with humans. When the Internet asks a question without being prompted by anyone or anything, it will be as conscious as we are! As far as anyone knows, the Internet has yet to ask a question—or even answer one—without being prompted by humans.

What are the elements of consciousness? First, a conscious entity must be *self-referential*. It must be able to understand itself. Second, it must be curious—inquisitive for its own sake. Then it must be motivated—it must have goals. These are the sparks of intelligent life that would signify the birth of a new species. [Most of the 7 million humans currently living exhibit one or more of these signs of intelligent life].

The hardware and software of the Internet currently has none of these. The Internet cannot explain itself; has no inherent curiosity; and lacks a goal. Will these limitations exist, forever? Will we wake up one day to witness Sparks of Kurzweil's inflection point? And if we do, will it be too late?

References

Bellhouse, D. R., The Reverend Thomas Bayes: A Biography to Celebrate the Tercentenary of his Birth, Statistical Science (2004), 19, 1, pp. 3–43.

Bell, Melissa, "Sohab Athar's Tweets from the attack on Osama bin Laden—read them all below, May 2, 2011. http://www.washingtonpost.com/blogs/blogpost/post/sohaib-athar-tweeted-the-attack-on-osama-bin-laden–without-knowing-it/2011/05/02/AF4c9xXF_blog.html

Clifford, Stephanie, "Social Media Are Giving a Voice to Taste Buds", www.NYTimes.com, July 30, 2012.

Gross, Doug, "Google Street View adds 250,000 miles of roadway", Oct 11, 2012. http://www.cnn.com/2012/10/11/tech/web/google-maps-street-view/index.html

Menczer, Filippo, Tracking the Diffusion of Ideas in Social Media. Meeting of the American Association for the Advancement of Science, Vancouver, Feb. 18, 2012.

Wesolowski A, Eagle N, Noor AM, Snow RW, Buckee CO (2012) Heterogeneous Mobile Phone Ownership and Usage Patterns in Kenya. PLoS ONE 7(4): e35319. doi:10.1371/journal.pone.0035319

Booms

4

Abstract
One of the most common forms of self-organization observed in the wild is called preferential attachment. But it is not a new idea. Rather, preferential attachment is a restatement of Gause's competitive exclusion principle, which says that in every ecosystem, only one species comes to dominate. Over time, one species gains a small advantage over the others, and marshals this advantage to a dominant or exclusive position. A monopoly forms. This fundamental law of complex systems explains the rise of monopolies, the emergence of structure in the autonomous system level Internet, and clustering of membership in online social networks. In fact, it may determine global ownership of the Internet over the next several decades. In addition, the diameter of an online social network decreases with increasing density of links (percolation), which may explain how and why the Internet contributes to the rapid spread of political unrest across the globe. As online social networks form and rewire themselves into highly structured networks, they also become smaller. Thus, online social networks like Facebook and Twitter evolve from random-like networks into highly structured, scale-free, small worlds. This is the source of their speed and power to influence people.

4.1 Not So Natural Monopolies

When angel investors Gardiner Hubbard and Thomas Sanders agreed to finance a young inventor named Alexander Graham Bell in 1875 they had no idea that the company they founded would one day rule the Internet—or what passed for the first Internet back in 1877. The business was quite simple, actually, and followed a proven model pioneered by John D. Rockefeller's Standard Oil Corporation, Andrew Carnegie's steel empire, Cornelius Vanderbilt's railroad empire, and other barons of the era. *Vertically* integrated AT&T tried to own and operate the entire value chain from manufacturing and selling handsets, switching equipment, and telephone wires to local telephone exchanges. This strategy made perfect sense because a vertically integrated enterprise was impervious to competition, and maximized profits.

Together the three men formed what we know of today as AT&T—the first company to sell voice communication services to consumers. Voice networks did away with telegraph operators and put the consumer in change. In today's business terminology, we say the trained telegraph operator was *disintermediated*—cut out of the middle. AT&T would disintermediate the middleman again with the automated direct-dial system deployed in 1919, thus eliminating the exchange operator (the transition to direct-dial was completed in 1978). The born-again company would repeat this strategy yet again with a leap into wireless telephony. It seems a good company cannot be kept down.

But AT&T's success did *not* carve a straight-line path to dominance. When Alexander Graham Bell's original patents expired, the company suddenly had 6,000 competitors! Theodore Vail (1845–1920) took over the company and came up with a solution that made AT&T one of the most successful and famous companies in the history of business. Under Vail's leadership, AT&T built a superior telephone network, extended it across the entire US, leased access to competitors to force interoperability with the Bell system, and then convinced the government that vertical monopolies like AT&T were good for the country. AT&T became one of the first *natural monopolies* of the modern era.

Vail was no philanthropist. His business strategy became known as *Vailism*—a method of using closed systems, centralized power, and as much network control as possible, to maintain monopoly power. [Vail's closed systems were not unlike Steve Job's (Apple Inc.) *walled garden strategy* of today—again with telephones]. Even President Woodrow Wilson was persuaded that "one system, one policy, and universal service" was in the best interest of the nation. The closed system survived and thrived through several challenging shifts in technology and politics. It was a resilient strategy that seems old-fashioned, today, but in fact is a universal law of competition.

By 1911 the company had become what Vail dreamed it would—a powerful monopoly under one system and one policy—a universally accessible communication infrastructure spanning the nation. In fact, the company became so powerful that the Department of Justice threatened to intercede many times over nearly 100 years of punctuated regulation. In a settlement called the *Kingsbury Commitment of 1913*, AT&T divested itself of Western Electric (the equipment manufacturing arm) and agreed to coexist with local independents. Western Electric was free to make equipment for AT&T and its competitors. But by 1924 AT&T had bought up 223 of the 234 rivals! AT&T was once again a natural monopoly.

The *Telecommunications Act of 1934* allowed AT&T to continue to operate as a monopoly until it was replaced by the *Telecommunications Act of 1996*. But the Department of Justice continued its attack on the communications giant. It took legal action against AT&T for a full decade, 1974–1984, resulting in dividing

AT&T into seven "baby bells", as they became known. But once again, the baby bells merged into 3 major companies by 2006: Qwest, Verizon, and AT&T. By 2013, AT&T was threatening to buy rivals like T-Mobile and Vodafone. It seems that the communications infrastructure wanted to become a monopoly regardless of competitive and political pressures.

AT&T has been dismantled three times over the past 130 years, only to return as a monopoly or near-monopoly each time. The company has experienced three major booms—the first started in the early 1900s and ended with the 1934 regulation; the second started in the 1930s and ended in 1974–1984 with the baby Bells break-up, and the third started in the 1980s and ended in 1996. Is it working on its fourth boom? After being dismantled for the third time in the 1990s, AT&T regrouped and by 2012 it was the eleventh largest company in the US. And AT&T is a major player in the international Internet infrastructure race—a competition amongst global corporations to dominate the Internet. In 2012, the "most connected" Tier-1 Internet Service Providers (ISPs) were Cogent/PSI, Level-3 Communications Inc, and AT&T Services, Inc. It seems that AT&T keeps coming back as a dominant—monopolistic—organization. Why?

The answer lies in a fundamental fact of nature: species like AT&T are the product of a form of *emergence* known as *preferential attachment*. That is, all complex systems like the Internet, power grid, or national economy emerge from seemingly unstructured or chaotic circumstances into structured dominant organisms. Unlike a planned building constructed by following a blueprint, complex systems like AT&T *emerge* without a blueprint or central planning. Chaos is turned into order. And order is turned into a booming business.

Emergence has been described by a variety of scholars as *spontaneous order*, *self-organization*, or the formation of global structure out of local modifications or mutations. AT&T emerged from the chaotic disorder created by thousands of independent operators, changing regulations, and fast-moving technology trends. Its repeated transformation from insignificant player to dominant player was a consequence of economic, political, and social forces in its environment. These forces continue to shape the communications infrastructure underlying the Internet and cell phone networks, today. And, AT&T is likely to emerge as a global monopoly just as it has in the past.

The question is, "why does emergence seemingly produce monopolies?"

4.2 Gause's Law

A Russian biologist, Georgii Frantsevich Gause (1910–1986) explained why infrastructure companies like AT&T repeatedly rise to monopolistic positions regardless of obstacles like regulation and competition. Gause discovered what he called the *competitive exclusion principle* in 1932. The principle asserts that in the long run, no two species within an ecological niche can coexist forever. When two species compete, one will be slightly more efficient than the other and will reproduce at a higher rate as a result. The slightly more capable species crowds out the weaker species. The less efficient and weaker species is doomed to marginalization or eventual extinction.

Competitive exclusion happens all the time in business. For example, Apple Inc and Microsoft Inc started businesses at about the same time and in the same ecological niche—personal computers. In the very early years of personal computers consumers had a choice of operating systems. A PC could run the Digital Research CPM operating system, the UCSD P-System, or the MS-DOS system. Microsoft gained a slight advantage over all of its competitors when IBM selected MS-DOS for its very successful personal computer. Microsoft leveraged this relatively small advantage to "reproduce at a higher rate" than Apple. By the late 1990s Microsoft was a monopoly with over 90 % of the personal computer operating system market, and Apple was on the verge of extinction. [Apple only recovered after moving into a new niche not dominated by Microsoft].

4.3 Self-Organization

The competitive exclusion principle is the biological equivalent of a more general principle known as *increasing returns* in economics, the *network effect* in marketing, and *preferential attachment* amongst network scientists. These are different terms for the same thing—*Gause's Law*. The idea is simple, but powerful, because when it is properly executed, relatively small advantages can be marshaled into major dominance. Microsoft, AT&T, and other infrastructure companies rose to monopolies by leveraging small preferences. Initially insignificant, small preferences compound growth and increase the competitive fitness of the winner. Preferential attachment is the most fundamental mechanism underlying most forms of emergence. Here is how it works:

Imagine consumers are nodes in a graph as shown in Fig. 4.1a. Square nodes are competitors and round nodes are consumers. Initially, the field of competitors in the disorganized nascent market of Fig. 4.1 consists of five competitors, each commanding roughly the same percentage of consumers. Market share is defined as the percentage of consumers attached to each competitor by a link. When a consumer selects a product from a competitor a link is established between the consumer and the competitor. Links are added according to preferential attachment.

Preferential attachment works as follows: consumers select competitors with probability proportional to market share. So, if competitor A has twice the market share of competitor B, then new customers are twice as likely to select competitor A over B. This selection process is repeated until all consumers are linked to one competitor. As time passes, all consumers eventually link up with one or more competitors. Preferential attachment explains why most people preferred MS-DOS for their personal computer and *iPhones* for their personal cell phone. It explains why Microsoft's market share boomed in the 1990s, and Apple Inc boomed in the late 2000s.

4.3 Self-Organization

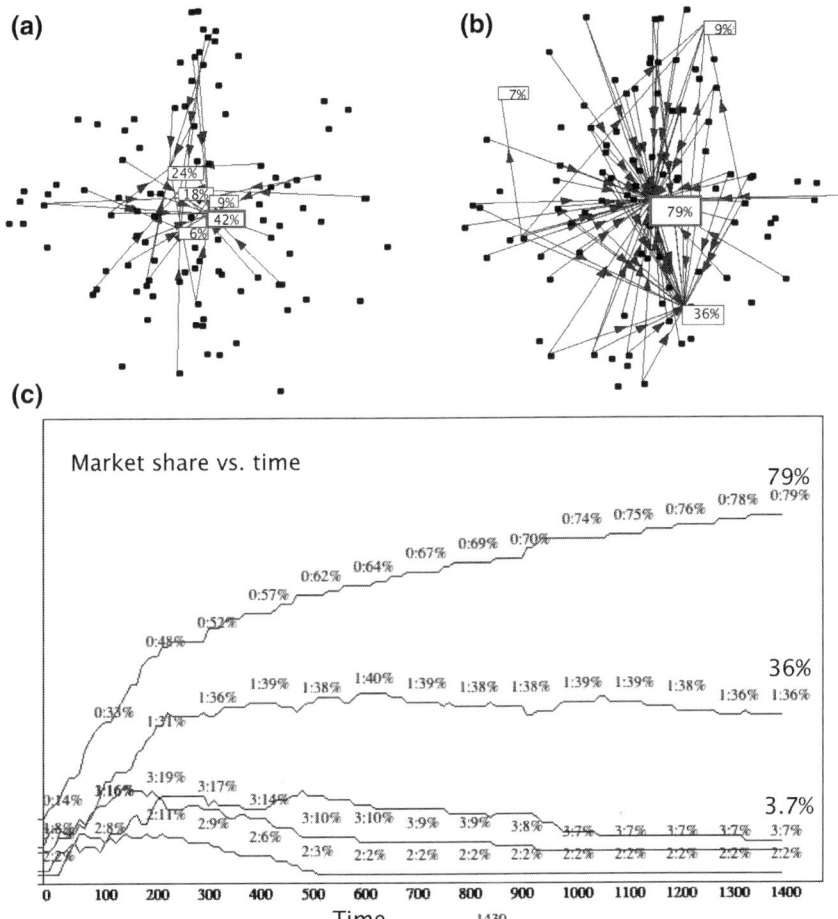

Fig. 4.1 A monopolistic hub emerges from an evolving nascent market because of *preferential attachment*. **a** *Square nodes* are competitors and *round* (*black*) *dots* represent consumers. **b** Eventually one competitor gains market share over all others and its increase in market shares accelerates. **c** Market share growth versus time shows how one dominant species emerges from the pack along an S-shaped adoption curve

Figure 4.1b shows what happens when the consumer-competitor network continues to evolve until all consumers are linked to a competitor. Figure 4.1c shows market share for each competitor versus time as the network self-organizes around the simple preferential attachment rule. A highly connected node—the *hub*—emerges with roughly 79 % market share. This is Gause's law in action. It is also an illustration of how structure emerges from a simple rule. At some point in the evolution one of the competitors achieves a market share (number of links) higher than all others. This small advantage is leveraged going forward, so that the market-share leader gains even more market share. In other words, market share begets more market share—and boom!

Each time the simulation of Fig. 4.1 is repeated, a different competitor emerges as the hub. The initial advantage that one competitor has over all others is purely accidental. Microsoft's dominance over all personal computer competitors during the 1980s was a combination of hard work and accidental gains in market share. Microsoft's competitors could just as likely have emerged as the monopoly that Microsoft became. Regardless, any advantage gives the lucky competitor a slightly better chance of acquiring more customers—just enough to start down a path of dominance.

This is also an example of *self-organization*. A complex system self-organizes when it adapts to its environment. Typically, self-organization means the system evolves from disorder to order. In Fig. 4.1, the market for personal computer operating system software evolved from a random network to a structured network with a hub at its center. Rather than the result of a carefully crafted plan, most self-organization is accidental and therefore unpredictable in advance. [But the distribution of market share among competitors follows a long-tailed power law!].

It is quite intriguing that a disorderly network of randomly connected consumer-competitor pairs organizes itself into a structured network with a hub. Typically, preferential attachment produces a similar hub structure every time the process is repeated. The hub always dominates with more than 70 % of all connections, while the other competitors settle for much less. And a competitor with 70 % or more is declared a monopoly. Are monopolies *always* a byproduct of self-organization? According to Gause the competitive exclusion principle always produces a monopoly.

4.4 Diffusion or Emergence?

Many scholars have studied the S-shaped curve describing the rate of hub emergence shown in Fig. 4.1 and explained by Gause's law over the past 150 years. One of the most remarkable insights linking emergence to the real world was the pioneering work of Frank Bass (1926–2006) during the 1960s and 1970s. He created the *Bass diffusion model* for describing product adoption rates. For example, his S-shaped model accurately predicted the widespread adoption of the personal computer and other products familiar to 20th century consumers.

Frank was born and raised in a small Texas town and he never really left it, although he held teaching positions at Purdue University. He was a long-time professor of management science at the University of Texas—Dallas, and received numerous awards for his contributions to mathematical marketing sciences. His 1969 paper, "A new product growth model for consumer durables" is so widely admired that it was recognized as one of the ten most frequently cited papers in the 50-year history of *Management Science*.

Preferential attachment verifies the Bass diffusion equation even though it is based on completely different assumptions. For example, Bass had no knowledge of Gause's Law. But is diffusion really what goes on when a product like the *iPhone* or personal computer rapidly gains widespread adoption by millions of consumers? Diffusion conjures up an image of molecules bumping into one another as one gas permeates another. Diffusion is an epidemiological phenomenon, while preferential attachment is a complex adaptive system emergence phenomenon. When a business booms onto the market, is it *diffusion* or is it *combustion*?

Bass's equation tells us how fast and widespread a product penetrates a market, while emergence tells us why only one competitor wins and the rest lose the competition for consumers. Diffusion ignores the winners and losers, while emergence ignores the underlying marketing mechanisms such as word-of-mouth, TV and radio advertising, etc. required to capture people's attention. You need both theories to explain booming monopolies.

The simulation in Fig. 4.1 ties these two models together. Emergence and Bass's diffusion equation produce identical adoption curves, but for different reasons. Emergence happens because of Gause's law and diffusion happens because of advertising and marketing. Emergence sorts out competitors, while diffusion ignores them. In any case, a business or organization booms when the product, service, or idea spreads faster than anyone expects.

The Internet and booming World Wide Web is no exception.

4.5 Internet Hubs

Is the Internet subject to Gause's competitive exclusion principle? After all, it is a fundamental infrastructure, and as illustrated by the history of AT&T, fundamental infrastructure companies tend to organize themselves into structured consumer-competitor networks with a monopolistic hub. It is too early to say for sure, but there is evidence that the online world is evolving toward hub-like control by a handful of dominant—perhaps even monopolistic—enterprises. By taking advantage of consumer preferences for one product, service, or idea over another, a handful of 21st century monopolies may be in the making as I write this chapter. Gause's law suggests that the next telecommunications giant will be at the Internet's center.

As evidence of self-organization that may be taking place in the Internet communications system, consider the vast wired network we know of as the *physical layer Internet*. Due to the 1996 Telecommunications Act, communications companies are allowed to co-locate in large buildings full of switching equipment known as *telecom hotels*. These buildings contain a variety of voice, data, video, and email switching equipment as well as gateways into the major backbones of the global Internet. These buildings are where many *cloud-computing* systems reside, because it is economically efficient to amortize costs across a number of tenants. Centralization is an advantage that companies leverage to grow faster than their competitors.

Very large telecom hotels and their tenants form the so-called *tier-1 autonomous systems network*, or ASN for short. The Internet's ASN forms its backbone or most critical pathways through the global Internet. Every autonomous system plugged into the Internet is assigned an AS number. To discover where a major switching node is, and how connected it is to the Internet, merely Google it's AS number. For example, the ten most connected AS's known circa 2011 were[1]:

#Links	AS Number: autonomous system owner
2972	174: Cogent/PSI
2904	3356: LEVEL-3
2365	7018: AT&T Services, Inc.
1959	6939: Hurricane Electric
1946	701: MCI Communications
1696	9002: ReTN.net Autonomous
1496	3549: Global Crossing Ltd.
1367	209: Qwest Communications
1332	4323: TW telecom holdings
1183	1239: Sprint

Cogent/PSI is the most connected AS because it has the most *peering relations*—links to other AS nodes. It is a hub. As you might expect, control of the most highly connected nodes in the ASN goes a long way toward controlling the Internet. But there are other measures of size. For example, the AS with the most bandwidth is *Deutscher Commercial Internet Exchange*, located in Frankfurt, Germany. These AS's obey a long-tailed distribution in terms of bandwidth and number of connections. The top 20 % account for 80 % of all Internet traffic.

Preferential attachment is organizing the Internet into a hub–like structure where a few super tier-1 autonomous systems dominate the global communications network. Why? The answer is simple: Gause's competitive exclusion says that an ecosystem has room for only one dominant species. And the species that wins is the one that is able to leverage its relatively small advantage as fast as possible. This means the AS with the *most connections* and *fastest transmission links* will most likely dominate all others, given time.

Look at the top ten highly connected AS's again. AT&T is third in line. It wasn't even in the top 20 in 2005. The communications ecosystem has expanded to encompass the entire globe. Global expansion has opened the ecosystem to more competitors. New predators are filling this larger and more expanded ecosystem, each one seeking an edge that will allow it to emerge as the global hub.

---

[1] http://as-rank.caida.org

4.5 Internet Hubs

The players with an edge will eventually dominate. Can AT&T rise above its competitors? Only time will tell.

If bandwidth is the measure of fitness in the Internet ecosystem, then the top ten list of bandwidth AS's excludes all of the connection leaders, but includes a host of new species. These are the companies that operate the nerve centers of the 21st century networks that will run our factories, transportation systems, energy and power systems, food and agricultural production systems, stock markets, and social interactions. Bandwidth will replace natural resources as the determiner of wealth. But note that not a single US company can be found in the list of the ten largest bandwidth leaders[2]:

Autonomous system owner	Bandwidth (Gb)
Deutscher Commercial Internet Exchange	4029
Amsterdam Internet Exchange	1180
London Internet Exchange	869
Equinix Exchange	946
Moscow Internet Exchange	570
Ukrainian Internet Exchange Network	319
Japan Network Access Point	273
Netnod Internet Exchange in Sweden	204
Spain Internet Exchange	168
Neutral Internet eXchange of the Czech Republic	171

4.6 Monopolize Twitter

Formation of monopolistic companies, emergence of *flashmobs*, and sudden appearances of "birds of a feather" communities in social networks like *Facebook* and *Twitter* all self-organize because of preferential attachment. Ultimately, one actor node ends up with more connections than any others, while most actors have only a few connections. The central actor starts to monopolize the community around him or her. When this hub actor gains even a slight popularity advantage over all others, its "friend list" grows faster than everyone else's list.

Monopolization of social network communities within online networks like *Twitter* and *Facebook* is a kind of Gause's competitive exclusion rule in action. Online monopolies are typically about ideas instead of products or money, but the concept is the same. When they happen, social network communities form around

---

[2] http://en.wikipedia.org/wiki/List_of_Internet_exchange_points_by_size

an idea or movement according to preferential attachment. As a result, *scale-free* structure emerges, because the pattern of "friend" connections obeys a long-tailed distribution—most actors have relatively few friends, while a few actors have a relatively large number of friends. You can observe this by plotting the fraction of nodes with 1, 2, 3... links against number of links per node.

Twitter is a textbook example. Users construct social networks by following each other online. Messages—called *tweets*—are purposely kept short and to the point so that users tweet often. A tweet posted by one user automatically appears on all followers' Twitter page. This automatic update creates a chain of transactions that link users together. [Users are *actor nodes* and links are the tweets as shown in Fig. 4.2].

The scale-free structure that emerges as a result of preferential attachment is easily observed in Twitter, where "friend" communities coalesce around a popular hub person. A few users belong to more than one community, as shown by strands of links reaching into more than one cluster in Fig. 4.2. Twitter and Facebook have millions of users, but in reality, most users connect with a much smaller inner circle of friends, and "friend" perhaps fewer than 150 other people.

A *retweet* is a message that has been forwarded to other users. A series of retweets reverberates through the social network like a contagion. Tweets containing information on an identifiable topic such as an expression of support for a politician, admiration of a celebrity, or comment on a newsworthy headline become fodder for big data aggregators like Topsy.com. Tweets can also express a negative sentiment such as opposition to a governmental policy or action.

Twitter topics are roughly identified by a *hashtag*. By analyzing the network formed by the spread of retweet contagions on a specific hashtag, social scientists can study the impact of an idea or movement on an entire community. Theoretically, network analysis can identify the most influential tweeter and the impact that a handful of people have on the connected community.

Studies of hashtag activity mirror the rise and fall of products as shown in Fig. 4.1c, but the time-scale is very different, see Fig. 4.3. Instead of spanning years, hashtags last perhaps hours or a few days. Twitter and Facebook products are ideas and sentiments, not *iPads* and eBooks, but their adoption obeys the same S-shaped Bass curve of the physical world. They are just short-lived.

Retweet networks and follower networks are two different structures superimposed on Twitter. Think of the follower network as a highway and the retweet network as the people traveling the highways. If you want to know how to get from point A to point B, follow the highways. If you want to know why traffic is congested on one part of the highway and not on another part, listen to the travelers. Are they jammed in traffic because of a sporting event, traffic accident, or early-morning commute? It is important to identify the topics being retweeted on Twitter if you want to measure the power of an idea sweeping across the community.

Researchers have studied the impact or influence of social networks on vast populations such as the people of Egypt during the *Arab Spring* and movements in the US by people in the *American Tea Party*. For example, consider the

4.6 Monopolize Twitter

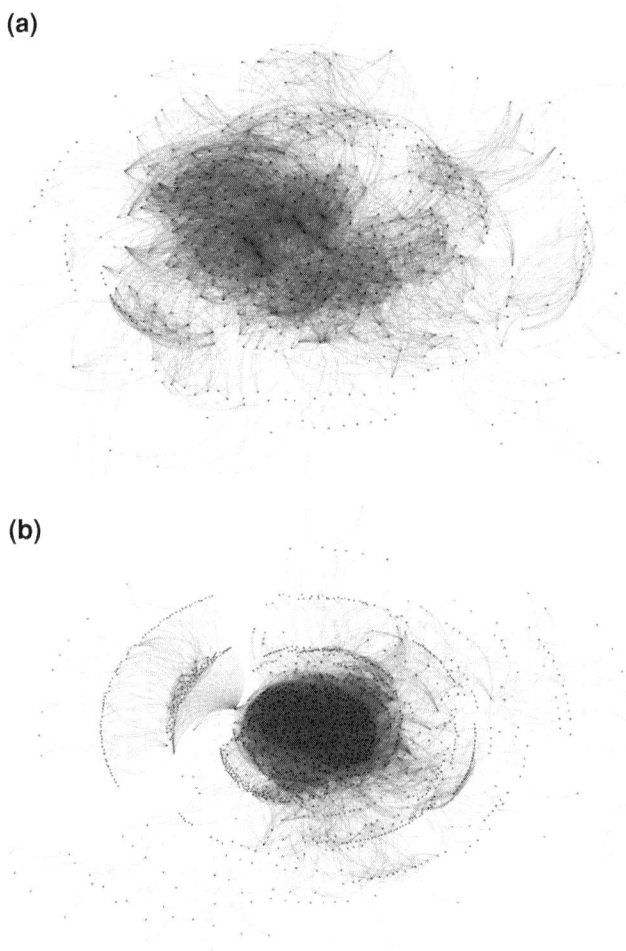

Fig. 4.2 Online social networks self-organize through a dynamic process of preferential attachment. Each cluster is a community surrounding a hub (most-connected user). Communities typically form around a popular user (a celebrity), idea, or friendships. **a** Online social network partially formed shows the emergence of clusters or tightly connected neighborhoods. **b** Same online social network shown in (**a**) after further evolution showing increased self-organization. A central core is surrounded by splinter groups with their own clustering

propagation of discontent spread through social media networks during the Arab Spring. The Arab Social Media Report issued by the Dubai School of Government says Facebook and Twitter were major tools used by Egyptians because they bridged the gap between social classes, and brought rich and poor together in a united front.[3] Nearly 9 in 10 Egyptians and Tunisians surveyed in March 2011 said

---

[3] Villareal (2011).

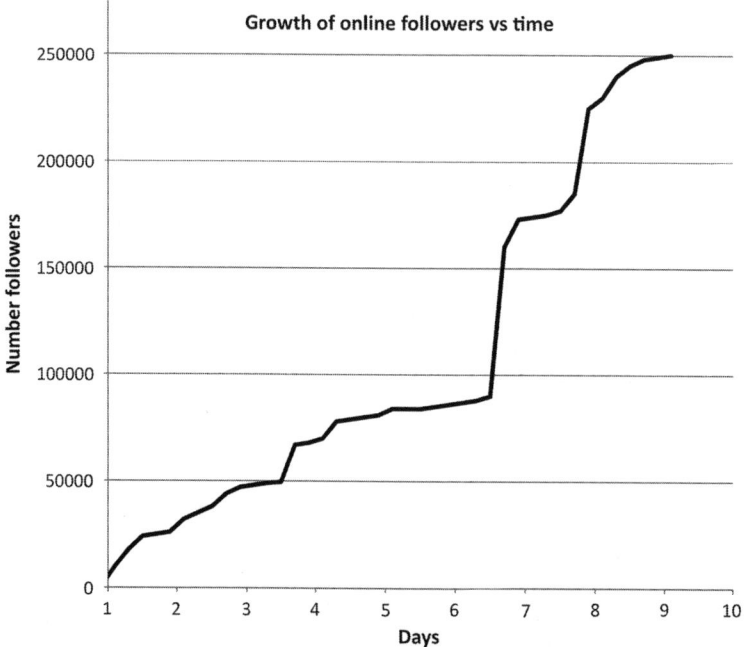

Fig. 4.3 Hashtag adoption rises and falls much like Bass's product adoption in the physical world, but the rise and fall happens much faster. Similar data can be found at: http://blog.ouseful.info/2012/02/09/what-is-the-potential-audience-size-for-a-hashtag-community/

they used Facebook to organize protests or spread awareness about their cause. Social media appears to be mobilizing, empowering, and shaping people's opinions.

How does social contagion work—is it diffusion or emergence? Like Google, Twitter uses *pagerank* (number of links to an actor node) to measure the "degree" of influence one actor has on another. Generally we know that the most connected user is the most influential person in a social network.[4] Community size obeys a long-tailed distribution—there are millions of small groups but few extremely large groups online. And so we think influence spreads like a human contagion— the larger the influence the more it spreads. For example, in a study done by Weng et al., researchers reported that 72 % of users follow more than 80 % of their followers, and 80.5 % are followed by 80 % of their followers.[5] In other words, there is a strong "follow-the-leader" urge among twitterers, and this translates into influence.

---

[4] Not always. Sometimes a connector actor is an intermediary that controls flow of information between major components of the social network.

[5] Weng et al. (2010).

4.7 Dunbar's Number

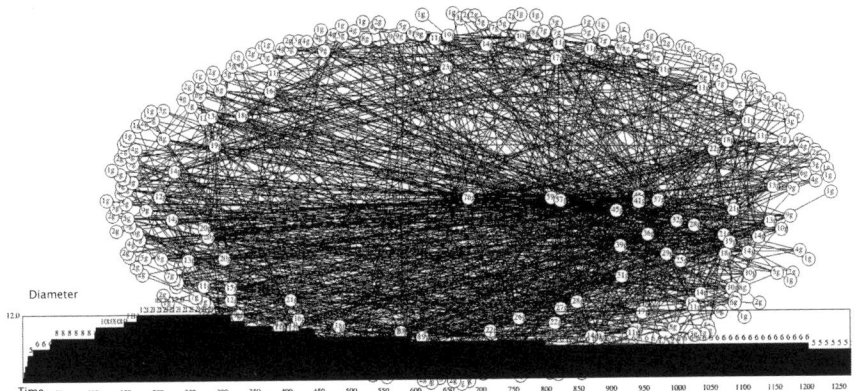

Fig. 4.4 The diameter of a scale-free network declines as more links are added, following an initial and brief rise during the early formation of the social network. Diameter of this 300-member network rises to 14 hops and then declines to 5 hops after 1200 links are added

4.7 Dunbar's Number

The science behind online social network formation is called *homophily*—the tendency of individuals to associate and bond with people just like them. Homophily and the confluence of sentiment, time, and preferential attachment combine to form scale-free structured online communities. This emergence may appear somewhat erratic and unpredictable but ultimately the resulting communities obey long-tailed distributions in both size and elapsed time between formations. What is the connection between homophily and network structure?

Two factors stand out: the emergence of a scale-free network structure with a dominant hub makes the distance between any two actors grow shorter as more connections are made, and the speed of message propagation increases as a handful of highly connected hubs form, see Fig. 4.4. Simply, birds-of-a-feather stick together, because of the small world property of scale-free networks.

We call the connectivity of an online friend the *degree* of that individual, and the strength of connectivity the network's *link density*. An actor's degree is equal to the number of links it has and the density of links is the ratio of links to total population of the networked community.

Figure 4.4 illustrates the importance of these two network factors on homophily. First, distance between actors is measured in *hops*, not miles or kilometers. If a tweet jumps from one person to another to get to a third person, we say it took two hops. The distance between two actors is equal to the number of hops separating them.

The chart along the bottom of Fig. 4.4 plots the maximum number of hops—called the *diameter*—required to send a message from anyone to anyone else in the network versus the total number of links in the entire network. As the number of links grows, the diameter shrinks, thus making the network smaller. [Initially the

Table 4.1 Effect of hub size and link density on social contagions

Link density	Random network	Scale-free network
Sparse (Avg. 8/actor)	Avg. 40 % actors reached	Avg. 62 % actors reached
Dense (Avg. 16/actor)	Avg. 90 % actors reached	Avg. 91 % actors reached

diameter rises, reaches a peak, and then declines]. But when the average number of connections per actor nears 7–8, the change in diameter levels off. The diameter of the 300-actor and 1200-link network of Fig. 4.4 levels off at 5. This means that a tweet can reach all actors in 5 or fewer hops. This is called the *small world effect* popularized by Stanley Milgram.[6] As the number of people we socialize with becomes denser, the (social) distance between us shrinks.

Homophily is facilitated by network percolation—adding links. Dense networks are smaller small worlds than sparse networks. Accordingly, dense networks are friendlier and easily facilitate homophily. Network structure and social homophily are intertwined.

The network of Fig. 4.4 is also scale-free, and scale-free networks are small worlds. It has one hub centered in the middle of the figure, and several other highly connected actors shown halfway toward the middle of the diagram. Most of the actors are only modestly connected. These are positioned around the circumference of the diagram. Hubs are important because they accelerate the spread of ideas. As the size of the largest hub increases, tweets and likes move even faster through the virtual crowd.

Not all messages get forwarded in a social network. Suppose only 20 % of them do. Then the fraction of messages actually reaching actors in the same network depends only on the infectiousness (20 %) of messages, the size of the network's hub (highly connected or not), the structure of the network (scale-free or random), and the density of the links (a lot of links per node or not). Table 4.1, below, summarizes these factors.

Assuming tweets are passed-on 20 % of the time, the number of people reached rises from 40 to 62 % simply by changing the structure of the network from random to scale-free and initiating the tweet from the network hub. But the difference between a structured and random network is nearly obliterated by link density. By doubling the density of links from an average of 8 links per actor to 16 links, 90 % of the 150 actors in this example are reached. Structure matters and so does link density. Surprisingly, social networks like Facebook and Twitter have self-organized to optimize the spread of messages. Even more surprising is the fact that they found an optimal structure on their own without central control or a blueprint.

What does this mean?

Hub size and link density determine the communicative power of a social network. They facilitate homophily.

---

[6] Lewis (2011).

4.7 Dunbar's Number

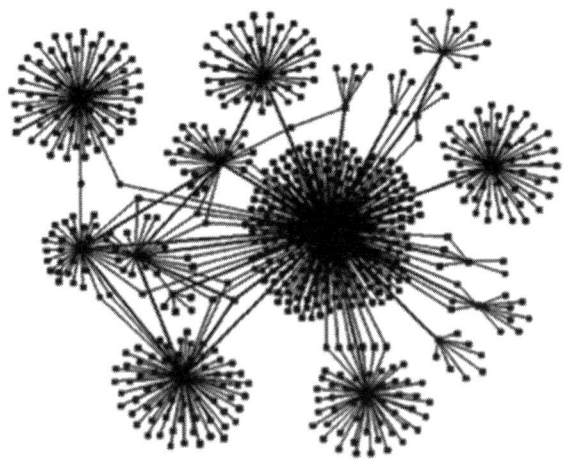

Fig. 4.5 The Internet has scale-free structure: major networks like Twitter and Facebook are filled with long-tailed scale-free galaxies and solar systems as illustrated here

During the Arab Spring, Facebook and Twitter actors wanted to get their message out, so a dense structure and the gravitas of a giant hub paid off. Ideas spread quickly and widely. But at some point the size of an organization slows things down. Or at least this is the theory of British anthropologist and evolutionary psychologist Robin Ian Dunbar (1947–). According to Dunbar the number of stable relationships anyone can maintain is approximately 150. This is *Dunbar's number*. Apparently more than 150 "friends" cognitively tax us, and fewer "friends" cognitively starve us.[7] If you believe in Dunbar's number then the ultimate size of a social network community is limited by our cognitive ability. Does this place an upper limit on social network communities? This is left as an exercise for the reader.

4.8 The Internet Is a Cauliflower

Most of the Internet is empty space—just like the cosmos. But it isn't an amorphous mass of wires, servers, web sites, and user communities. It isn't random nor is it chaotic. It has structure, and currently that structure looks like a head of cauliflower, see Fig. 4.5. Furthermore, it isn't an open system devoid of governmental oversight and freedom of use. The open Internet is a myth—a popular meme that has survived the early days before the Internet was commercialized.

By 1998 the Internet had matured to the point where it could be privatized. The NTIA (National Telecommunications and Information Administration within the US Department of Commerce) produced a "Green Paper" describing how the Internet should be governed, how to transition the domain name servers, DNS, to private ownership, proposed adding more gTLDs (global Top-Level Domains such as.tv for television), proposed that trademarks be honored as Internet names, reduced the $50 DNS registration fee to $20, and set aside 30 % of the revenues

---

[7] Dunbar (2010).

from DNS registration for the Intellectual Infrastructure Fund (IIF). The Green Paper established Internet Corporation for Assigned Names and Numbers (ICANN) to sell blocks of names to several authorized re-sellers.

Soon after its commercial launch in 1998 the Internet began evolving according to Gause's competitive exclusion principle. By 2012 Gause's law started to become visible. First came ownership of the Internet's bones—the cabling and server infrastructure that everything else depends on. Two companies in 2012 managed over 80 % of the bandwidth available online, and the top ten Tier-1 Internet Service Providers managed most of the peer-to-peer connections holding the Internet fabric together.

Second came governmental oversight. In fact, open Internet governance may be a moot topic of debate by the time you read this. Without cabling and servers, an open cyberspace becomes an empty ghost town. So more cabling and servers exist today than ever before. However, they are privately owned and operated. In fact, Internet governance has slipped out of the hands of volunteer committees and open standards-making organizations into the arms of corporations and governments regulating it. Take *net neutrality* as an example.

Professor Tim Wu of Columbia Law School popularized the term "*net neutrality*" in 2003. Wu argued that providers of Internet services should treat all Internet content, web sites, and devices, equally.[8] According to Wu, net neutrality is the equivalent of the First Amendment to the US Constitution, which guarantees freedom of expression. Since the Telecommunications Act of 1996, information transmission over publicly switched networks (the Internet and voice) has prevented communication companies from restricting content, types of users, types of equipment, or speed of information flow in any way.

For example, a communication company cannot charge more for a picture than email. Bits are bits—it does not matter what they represent. But this freedom seems to be slipping away. So-called value-added products are making it possible to charge more for some bits than others.

Net neutrality came to the fore when AT&T announced in 2005 that it would charge some customers more than others, depending on the company and type of content being transmitted over AT&T lines. Google.com and Amazon.com vigorously opposed the new billing algorithm, claiming that AT&T's prioritized treatment of data was discriminatory. In 2006 the Federal Communications Commission persuaded AT&T to back down for two and a half years following its acquisition of BellSouth. Then, in 2012, AT&T announced it would *throttle*[9] bandwidth to heavy users, thus discriminating according to how much load a user puts on their network. As I write this, throttling is common practice among Internet service providers, and was backed up by a US Federal court. As of 2014, net neutrality is dead—a decision that is likely to accelerate Gause's competitive exclusion principle.

---

[8] Wu (2003).

[9] *Throttling* is the process of slowing down transmission to avoid traffic overload.

And then there is *real governance*—by *real countries*, not virtual communities. In 1998 the US Department of Commerce outsourced all technical, policy, and operational details to private companies. Allocation of names and IP numbers was outsourced to the *Internet Corporation for Assigned Names and Numbers* (ICANN), headquartered in Marina del Rey, California. But, more recently the Internet has attracted the attention of a much bigger governing body—the *United Nations*.

ICANN allocated segments of IP numbers to five Regional Internet Registries (RIR) spread across the entire planet. ICANN control is delegated to these five regions, but who controls ICANN?

AfriNIC	African Network Information Centre (Africa)
APNIC	Asia-Pacific Network Information Centre (Asia, Australia, New Zealand, and neighbors)
ARIN	American Registry for Internet Numbers (United States, Canada, parts of the Caribbean, and Antarctica)
LACNIC	Latin America and Caribbean Network Information Centre (Latin America and parts of the Caribbean)
RIPE	Réseaux IP Européens Network Coordination Centre (Europe, Middle East, and Central Asia).

In 2006, the United Nations organized the Internet Governance Forum (IGF) to address policy issues that were previously the domain of Internet volunteers and pioneers. This governing body is still evolving, but if preferential attachment takes over, it is only a matter of time before IGF becomes a governing hub. The old open system principles that built the Internet are unlikely to survive global socio-political competitive exclusion. ICANN may hold the monopoly now, but a new species is on the prowl.

At the very highest level, Internet governance is being molded into a cauliflower-like structure, too. This is the consequence of preferential attachment and Gause's competitive exclusion principle operating on a global socio-political scale. Who owns the Internet?

References

Dunbar, Robin I. M. (2010). How many friends does one person need?: Dunbar's number and other evolutionary quirks. London: Faber and Faber. ISBN 0-571-25342-3.
Lewis, Ted G. (2011), Bak's Sand Pile, AgilePress, 2011, 378 pp.
Villareal, A. (2011), Social Media a Critical Tool for Middle East Protesters, VOA News.com, March 1. http://www.voanews.com/english/news/middle-east/Social-Media-a-Critical-Tool-for- Middle-East-Protesters-117202583.html.protesters-117202583/172762.html.
Weng J., E. Lim, J. Jiang, and Q. He (2010), "Twitterrank: Finding topic-sensitive influential twitterers," in Proceedings of Third ACM International Conference on Web Search and Data Mining (WSDM), pages 261-270, 2010.
Wu, Tim, (2003), "Network Neutrality, Broadband Discrimination", Journal of Telecommunications and High Technology Law 2: 141. doi:10.2139/ssrn.388863. SSRN 388863.

Bubbles

5

Abstract

What causes bubbles? Why are complex systems like the national economy punctuated—populated by long periods of steady growth (or decline) followed by excesses leading to a crash? The answer lies in a deeper understanding of complexity and Rosenzeig's Paradox of Enrichment. When the economy is going along just right, speculators jump in and enrich one or more segments of the economy. They over do it, and guess what? That segment of the economy collapses. It's called a Minsky moment, Asian Contagion, Subprime mortgage collapse, and other names, but it always comes back to Rosenzeig's Paradox of Enrichment. Perhaps the most dramatic illustration of this systemic effect is the financial meltdown of 2008, which saw the rise and fall of the overheated real estate market in the US. When homeownership exceeded its carrying capacity of approximately 65 %, the housing segment became nonlinear and collapsed. The punctuation was so widespread that it brought down the entire economy.

5.1 Financial Surfer

My friend Dave is a financial surfer. During the run-up of the Internet dotcom startups of the late 1990s, he speculated on Pet.com, AOL.com, and Kozmo.com. As the dotcom wave became a tsunami, Dave's investments also grew bigger and bigger until they reached a peak in March 2000, and then popped like bubblegum on a pre-teen. Dave was fortunate—he sold his equities a year earlier when they were near their peak values.

History is littered with stories of bubbles and the financial surfers who rode them. Perhaps the most famous bubble was the tulip craze in 1630s Holland. At the peak of tulip mania, in February 1637, a single tulip bulb sold for 10 times the annual salary of the average Dutchman. Then there was the 1720 South Sea

Bubble, the Mississippi Bubble (1719–1720), the Bull Market of the Roaring Twenties (1924–1929); and Japan's "Bubble Economy" of the 1980s. These have become so commonplace they have a name—*Minsky moment*. Wikipedia describes a *Minsky moment* as the point where exuberant and over-indebted investors are forced to sell assets to pay back their loans, causing sharp declines in financial markets and jumps in demand for cash.[1]

Paul McCulley of PIMCO used the colorful phrase *Minsky moment* to describe the 1998 Russian financial crisis. Inflation reached 84 % in August 1998 forcing the Russian republic to default on loans. Banks closed, people lost their life savings, miners went on strike, and protests organized across the country. The ruble fell from $0.179 to $0.048—a 75 % plunge. The Russian Minsky moment followed a period of prosperity and increasing values, which drove speculators to spend borrowed money chasing ever-increasing investments. When those investments sharply fell in value, the balloon burst. In the case of the Russian Minsky moment, the bottom (briefly) fell out of the price of oil—an economic mainstay of the country.

Hyman Minsky (1919–1996) first proposed the "The Financial Instability Hypothesis" to explain the vicious cycle that bears his name.[2] The theory says that in any credit or business cycle investors eventually reach a point where they have cash flow problems due to the spiraling debt incurred by irrationally exuberant speculation. At the tipping point there are no buyers for the elevated stock, real estate, bond, or business, because the price is simply too high. When the peak is reached, a major sell-off begins leading to a sudden and precipitous collapse in prices and a sharp drop in market liquidity. Minsky moments are the point at which bubbles pop. They are caused by irrational and extreme excess.

5.2 Worthless Homes

My friend Dave rode the infamous Housing Bubble that began on the heels of the dotcom bust circa 2001–2006. In 2001 the Federal Reserve began lowering interest rates to rescue the economy from the ravages of the dotcom Minsky moment. Homeowners were paying rates over 8 % on a 30-year mortgage in 2001. This relatively high rate steadily declined to 3.8 % by early 2012 and then 2.75 % in 2013 when this chapter was written. Monthly payments on a $200,000 loan in 2001 plunged from $1,500/month to $816/month in 2013. In other words, the cost of borrowing money dropped by almost one-half in less than a decade. To Dave, this was an invitation to borrow all the money he could.

In 2003, Dave started surfing once again. Only this time he rode the wave of cheap money. First, he bought a large house that quickly appreciated by 10 % in one year! Dave sold it and used his profit to buy two houses. He lived in one and

---

[1] Lahart and Justin (2007).

[2] Minsky and Hyman (1992).

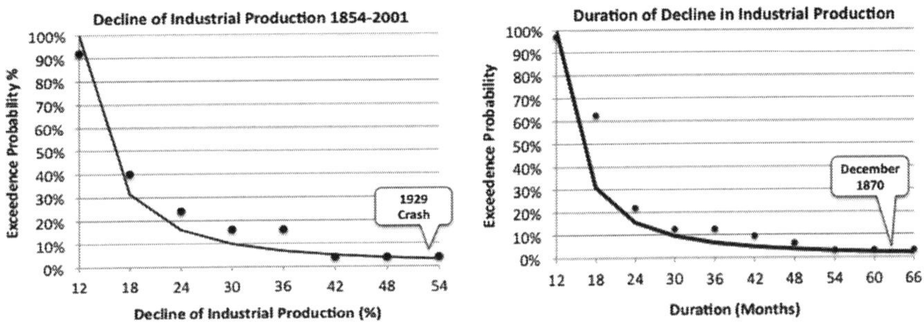

5.2 Worthless Homes

Fig. 5.1 The sizes and durations of financial bubbles within the US over the past 150 years obey a long tailed distribution. The 1929 crash was the deepest and the 1870 crash was the longest

remodeled the other one, and sold it for a 10 % profit. This was easy money, so Dave expanded. Each time he bought another property, he made a minimum down payment and leveraged the investment as much as possible. By 2007 Dave held titles to five houses worth more than $4 million. At this rate his highly leveraged investments would make him a very rich man in a few short years.

Dave's logic was watertight. In America, nobody defaulted on a mortgage. After all, people have to live somewhere. And, homeownership was the best investment a family could make. Owning a house was better than putting your money under the mattress, buying risky stocks and bonds, or investing in the children's college fund. That was the popular *meme* permeating the financial ether back in 2007. And nearly everyone drank the cool-aid.

But then the inevitable happened. The housing market collapsed, Dave's investments dropped in value by 45 %, and many of his neighbors simply abandoned their houses because they owed more on the mortgage than the house was worth. The housing bubble popped in one gigantic Minsky moment. Dave tried to hang on for another 18 months, but by 2012 he was at the end of his rope. His $4 million nest egg suddenly became worth $2.25 million. Even worse—Dave owned $3.6 million! His profit suddenly became a debit of $1.35 million.

In California surfer terms, it was a wipeout, and Dave went under.

History is full of stories like this. Bubbles happen all of the time. Some of them are huge like the 2001–2008 housing bubble, but most of them are small. Bubbles like this are long-tailed events. See Fig. 5.1. Financial bubbles follow Levy Flights in terms of size and duration. The next one of size similar to the 2008 financial meltdown should happen again in about 24 years.

But enough about Dave. The real question is, "why?" Why do financial bubbles happen in the first place, and what can we do about them in the second place? Surprisingly, we know the answer, because waves, surges, and bubbles happen all the time in nature. They are self-similar fractals.

Nature holds the key to understanding this kind of fractal.

5.3 The Paradox of Enrichment

In short, bubbles are caused by having too much of a good thing. This phenomenon was observed and studied scientifically by Michael Rosenzweig (1941-) in 1971. Dr. Rosenzweig is a noted ecologist and founder of the Department of Ecology and Evolutionary Biology at the University of Arizona. Wikipedia succinctly describes what Rosenzweig called the *Paradox of Enrichment*, "A common example is that if the food supply of a prey such as a rabbit is overabundant, its population will grow unbounded and cause the predator population (such as a lynx) to grow unsustainably large. This may result in a crash in the population of the predators and possibly lead to local eradication or even species extinction". By enriching the food supply (prey) of a predator, it is possible to kill both predator and prey.

Rosenzweig's paradox is a paradox because it says that enriching an infrastructure such as air, water, food, or prey produces fewer predators rather than more. Shouldn't an abundance of food stimulate more growth and more abundance? Quite the opposite—making the ecosystem "richer" damages it! But not always.

In the world of housing speculation, think of the predator as the homeowner/buyer. The prey is the lending agency or source of mortgage money. Enrichment came in the form of cheap money. Specifically, the Federal Reserve made cheap money abundant and available by lowering interest rates and qualification of borrowers. As a result, speculation spiraled out of control. Banks enriched themselves with cheap money and borrowers enriched themselves with low-interest loans and rising resell values.

According to Rosenzweig's theory, enrichment eventually reaches a tipping point where the *carrying capacity* of the predator-prey system is exceeded and suddenly the ecosystem collapses. Both predator and prey overshoot, which introduces an instability that rocks the boat for the entire ecosystem. In the housing ecosystem, carrying capacity is the capacity of an economy to support a certain level of borrowing. As I will show later, the carrying capacity of the US housing market can be measured by the percentage of *homeownership*.

So we have three factors to consider here: the predator, prey, and carrying capacity of the ecosystem. The predator is anything or anyone that consumes a prey; the prey is anything or anyone that is consumed; and carrying capacity is anything or anyone that directly supports the predator-prey ecosystem. That is, an underlying infrastructure exists with limits on how big, robust, or enriched it can be, and still not destabilize the ecosystem.

In this case, the carrying capacity of the housing market was exceeded by the general economic strength of the country—GDP (Gross Domestic Product). In effect, the housing bubble burst because the US GDP was unable to support the rapid increase in debt burden assumed by borrowers. The money supply enriched this ecosystem by expanding too rapidly when the Federal Reserve artificially lowered interest rates and printed money. This rapid expansion sent shocks through the financial system, destabilizing it, and ruining the very ecosystem it was supposed to save.

5.4 Housing State-Space

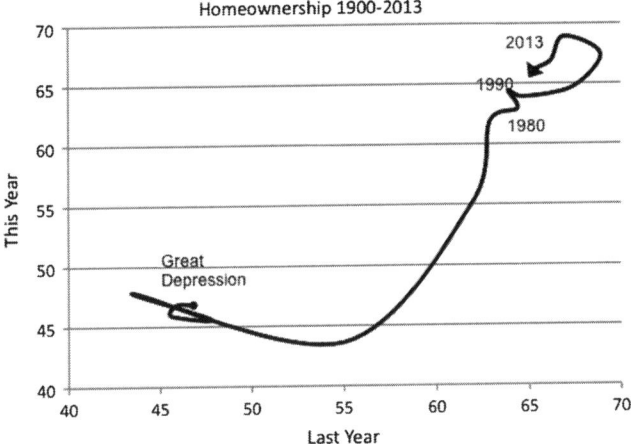

Fig. 5.2 State-space diagram of homeownership over the past 90 years. Both *horizontal* and *vertical axes* are in percentages of eligible homeowners. Nonlinear tipping points are shown during the Great Depression, 1987 and 2008

5.4 Housing State-Space

Does paradox of homeownership enrichment hold upon deeper investigation? Figure 5.2 shows the state-space diagram for homeownership since 1900. A *state-space diagram* is simply a plot of the state of a system as it moves through time. The states in this case are homeownership in year (t) versus year (t − 1). That is, homeownership, as a percentage of the population of eligible owners, is plotted against itself. Ownership increases and decreases over time, but tends to stabilize or not, depending on the paradox of enrichment principle.

State-space diagrams are useful for detecting trends in *time-series data*. For example, if a plot of the system's state forms a straight line, the trend is obviously in one direction (up or down). If the plotted states form a circle, ownership oscillates around some *fixed point*. A fixed point is a state that attracts the system by pulling it toward the fixed point. If the cycle goes to zero, the system dies! Dying out is equivalent to a species going extinct in the natural world. So we have the following state-space diagram patterns to consider:

Trending up/down: A line that goes up or down.
Oscillating but stable: an elliptical or circular pattern.
Collapse: a line that ends at the origin or (0,0) point in the diagram.
Chaotic: erratic and wildly gyrating pattern.

So what does the state-space diagram of Fig. 5.2 say? Homeownership was unstable and oscillated around a fixed point of about 47 % during the Roaring 20 s, and then dropped precipitously during the Great Depression. Immediately after WWII ownership steadily increased along with Gross Domestic Product (GDP as carrying capacity) until the late 1980s. A real-estate instability in 1988–1991 caused a turbulent and chaotic period that appears as a small twist in the diagram. Then in 2005 the upward trend hit a wall. From 2005 until 2013 ownership began circling around another fixed point of approximately 65 %, after declining from a peak of almost 70 %. The carrying capacity of the US was apparently 65 % during the period 2005–2013, and seems to be circling this fixed point.

Naturally a curious mind wonders what forces shape the state-space diagram. Does homeownership depend on the weather, political winds, economics, or some hidden force? Of course homeownership is desirable, so why isn't it 100 %? Why is it 65 % instead of 47 % like it was almost 100 years ago? The answer lies in understanding economic carrying capacity—the ability of an economy to support a certain level of wellbeing or "life style."

Apparently, the carrying capacity of US homeownership is currently about 65 %—the contemporary fixed point. Increasing ownership beyond this level introduces instability due to enrichment of the prey—the homeowner in this case. And it is also a paradox, because making homes more affordable makes them less affordable! Why? The answer lies in Rosenzweig's biological complexity theory. As borrowed money became more plentiful and inexpensive the system became less able to support more buyers. And when the ownership rate soared above 65 %, borrowers in the US exceeded the carrying capacity of the (US financial) system. Enrichment killed the very thing it was designed to enhance.

5.5 US Economic Carrying Capacity

What evidence is there that the US economy cannot support a homeownership rate in excess of 65 %? To answer this question, and to dive deeper into the complexity of a seemingly simple system, consider Fig. 5.3. Homeownership-as-predator is plotted versus GDP-as-prey in this state-space diagram. That is, the state of the system is defined by both percentage of homeownership and GDP, combined. If we assume GDP is a suitable measure of wealth, then more people should be able to afford a house when wealth increases. But, if GDP surges, the economy risks overheating and carrying capacity is exceeded causing the economy to collapse. Strange as it may seem, it is possible for the economy to be too good.

My wealth hypothesis is not entirely supported by the state-space of Fig. 5.3, because GDP has steadily increased since 1947, with one exception—the 2008 financial meltdown. But homeownership began declining from 2005 and continues to decline as I write this chapter. [At the time I write this, 2014, I do not know where it will end]. The rate rapidly increased during the 1950s, leveled off through most of the rest of the 20th century, and then took off again until 2005. Then the rate of ownership plunged. Currently, GDP is slowly expanding, but ownership is not.

5.5 US Economic Carrying Capacity

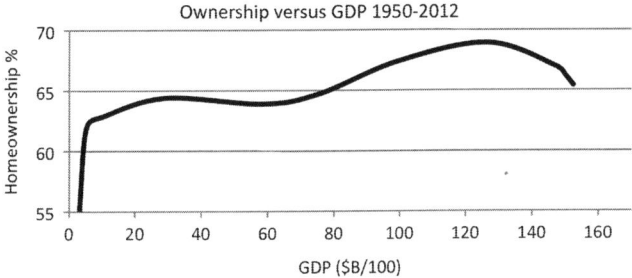

Fig. 5.3 Homeownership versus GDP (Gross Domestic Product), 1950–2011, peaked at nearly 70 % in 2005 and then declined to historical levels

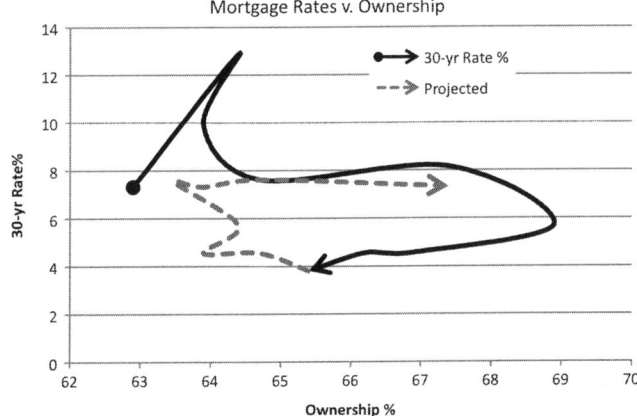

Fig. 5.4 State-space diagram of mortgage interest rates versus homeownership rates. Both *horizontal* and *vertical axes* are in percentages

Another piece of evidence is shown in Fig. 5.4. This state-space diagram compares the 30-year mortgage interest rate with homeownership rates. In the language of the ecologist, homebuyers prey on savings and loan banks, so lenders are a kind of prey. Once again, enrichment comes in the form of cheap and plentiful money. And during the decade leading up to the 2008 financial meltdown, almost anybody was allowed to purchase a home for little or no down payment, regardless of collateral or credit rating. Liberal lending requirements led to the infamous *subprime* mortgage debacle.

Starting in 1970, Fig. 5.4 shows interest rates shooting up to a maximum in excess of 18 % on October of 1981. Then the rate dramatically fell throughout the 1980s and 1990s as ownership steadily rose. The rate leveled off and then slightly rose once again to a peak of 8 % in 2000. The dotcom bubble collapsed in 2000, but the enrichment continued. In fact, the dotcom crash motivated the

Federal Reserve to lower rates even further as it simultaneously increased money supply—fomenting a financial double-whammy. Then the bottom fell out of housing because the homeownership carrying capacity was exceeded, and the system's state-space diagram began to circle around a fixed point of roughly 65 %.

The financial system's slow adaptation to a new fixed point is incomplete, so it is not clear where it will end. I have taken the liberty of projecting the state-space diagram into the future along the dotted line in Fig. 5.4. My projection suggests that future interest and ownership rates will return to historical levels. The circling corresponds with ups and downs in the business cycle, which center on traditional levels. This is pure speculation, of course, but if we believe in Rosenzweig's Paradox of Enrichment, the fixed point will remain fixed until the carrying capacity of the economy changes. [This would require a major contraction or expansion of GDP].

Carrying capacity is often an overlooked cause of many collapses of complex systems. In fact, it is rarely included as a consideration whenever politicians and policy makers create policy. Both Presidents Clinton and Bush declared homeownership a national priority with little appreciation or knowledge of the Paradox of Enrichment. Perhaps they skipped their biology classes in college. Nobody asked, "can the US support a higher rate of homeownership than we enjoyed in the 1990s?" Instead, policies were driven by the idea that, "More people should own their own homes." In 21st century reality, homeownership is a byproduct of a more complex and dynamic system. Regardless of politics, an economy can only support what its carrying capacity allows.

Interestingly, the actions of a few Americans (real estate speculators) led to an economic collapse for all Americans. Apparently, no man or woman is an island.

5.6 The Commons

The homeownership ecological disaster of the first decade of the 21st century is an example of what sometimes happens to a more general system called a *commons*—loosely defined as a resource shared among members of a community. For example, the Interstate Highway System, Fisheries, Forests, Social Security, Open Source Software, and Public Parks are well-known commons, because they encapsulate a resource shared by a community of users. A commons includes an infrastructure and the entities that depend on it.

The idea of a commons goes back hundreds, and perhaps even thousands of years. It is highly likely that our ancient ancestors formed various commons to raise herds of cattle, collectively farm shared land, and raise armies to protect the village from raiders. Today the air we breathe, the water we drink, the spectrum we use for radio and Internet communication, and the fuels we consume to produce electricity are various kinds of commons.

Ecologist Garrett Hardin (1915–2003) popularized the concept of a shared commons in 1968. His article in *Science* magazine titled "The Tragedy of the Commons", relates a parable described by William Forster Lloyd (1795–1852) in which villagers share a common parcel of pastureland, called the *commons*.[3] This commons worked as follows: privately owned cows were allowed to eat as much grass as they wanted from the community-owned commons. However, the cost of the commons was shared, equally. Everyone paid an equal amount to support the pasture. But here is the problem—individual villagers kept the profits derived from the sale of cows regardless of the amount of grass consumed by each cow. They exploited their fixed-costs (leverage) to make a variable amount of profit.

In Hardin's parable a rational villager was rewarded for putting as many cows out to pasture as possible, with little concern for the impact on the supply of grass. Each villager acted in his or her own self-interest, naturally. But maximizing self-interest leads to overgrazing, which leads to depletion of the grass, which leads to depletion of cows. And, with no cows to sell, the villagers also became depleted. By increasing each villager's herd to maximize profits, the villagers destroy their cows, and therefore, themselves. What started out as a good idea ended in collapse of the shared commons.

The tragedy of the commons is simple to understand, and yet it happens all the time, because most policy-makers never consider the carrying capacity of a complex system. Today's shared resources, such as the Interstate highway network, air we breath, and water we drink are suffering from the tragedy of the commons. Why do people perform actions that ultimately lead to their ruin?

5.7 Simple but Complex Ecosystems

Hardin's commons is a simple system, but it can behave in complex ways. What is more elementary than predator cows and preyed-upon grass? Given the right rate of growth of the grass and a balanced load of cows it is possible to sustain the pasture forever. The predator and prey are in harmony. But if too many cows are put out to pasture, the cows deplete the grass and eventually themselves, too. Greed can make this predator-prey system go extinct. On the other hand, if too few cows are let onto the commons, grass goes to waste and profits suffer. In spite of its simplicity, it may be difficult to find a balance between the cow and grass population that simultaneously guarantees sustainability of the commons and maximum profit.

Simulations of the cow-and-grass commons illustrate the delicate balance required to run a sustainable commons. Figure 5.5 plots the dynamic change in cow population and grass level for two cases described above: (a) sustainable, and (b) unsustainable. When the cow and grass populations are in harmony, the amount of each may oscillate as shown in the upper plot of Fig. 5.5, but they never go extinct. On the other hand, when the ecosystem is out of balance, the entire system collapses.

---

[3] Hardin and Garrett (1968).

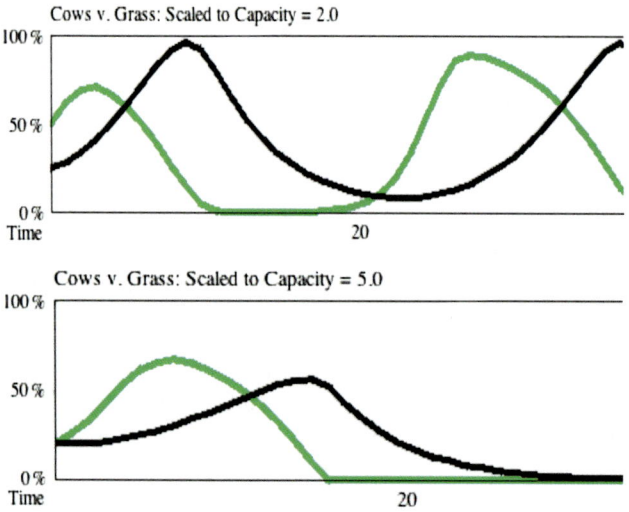

Fig. 5.5 Dynamics of the cow (*black line*) and grass (*green line*) population: the *upper* plot shows how cow and grass populations oscillate, but remain sustainable. The *lower* plot shows how enriching the cow population can destroy the entire commons

This ecosystem contains the seeds of its own collapse. It is susceptible to nonlinear distortions, because it has tipping points. The cow-grass system may operate for long periods of time without surprise. It may even tolerate a small amount of enrichment. But at some point, even a small increase in the rate of growth of cows or grass can cause the entire system to collapse as shown in the lower plot of Fig. 5.5.

For example, suppose the stable, sustainable, and cyclic system shown in the upper plot of Fig. 5.5 is slightly modified by adding a small amount of fertilizer to the grass to make it grow faster. If it is doing its job, fertilizer should increase the rate of growth of grass and in turn, allow for more cows. What actually happens comes as a surprise. The grass grows faster, more cows are added, and the commons promptly collapses! Why?

As you can see in Fig. 5.6, the enrichment of grass causes the cow population to overshoot the carrying capacity of the system. The extra cows quickly deplete all of the grass, which in turn depletes the cow population. In effect, providing more food to the cows ends up killing them. This happens because growth rates of grass and cows are out of sync. As these rates get further out of sync the system becomes less stable and oscillates. Eventually, the oscillations become chaotic and the system "blows up."

This dramatically illustrates the nonlinear nature of this simple system. It also illustrates a counter-intuitive explanation for bubbles. Like shared cow pastures, economies often overshoot their carrying capacities which eventually reach a nonlinear tipping point. When overshooting happens, the bubble bursts. Minsky moments are nothing more than the bubble that bursts following enrichment.

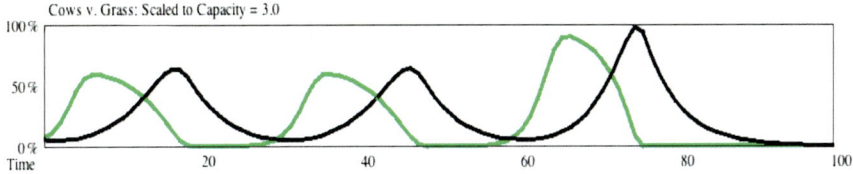

Fig. 5.6 Enrichment of the grass (*green line*) leads to its extinction, and in turn, the extinction of the predator cows (*black line*)

5.8 The Road to Oblivion

Is the parable of the commons simply a parable? Or does it relate to the 21st century? I claim that it not only relates to our contemporary society but it is an essential tool of 21st century policy-makers, politicians, and every-day business executives. Allow me to argue the point using contemporary examples of extremely critical commons that we depend on every day.

The US Interstate Highway system is one of the marvels of the 20th century. It has supported a vast and powerful economy, provided mobility for a nation, and remains the backbone of national security. Officially called the *Dwight D. Eisenhower National System of Interstate and Defense Highways*, it "is a network of limited-access roads, including freeways, highways, and expressways, forming part of the National Highway System of the United States. Authorized by the Federal Aid Highway Act of 1956, the network … has a total length of 47,182 miles. As of 2010, about one-quarter of all vehicle miles driven in the country use the Interstate system. The cost of construction has been estimated at $425 billion (in 2006 dollars), making it the largest public works program since the Pyramids".[4]

The Interstate highway system is a shared commons. Construction and maintenance is funded by a combination of federal and state gasoline taxes, tolls, taxes on trucks, and occasional supplemental funding from Congress. The Highway Trust Fund was established in 1956 to collect and distribute these funds. Freeways, and the land they rest on are owned by the states, while the Federal government dictates construction standards, route numbering, etc. In general, these funds are used to extend the network and maintain roads and bridges.

Is this commons subject to the complex dynamics of Hardin's commons? Is it possible for the Interstate Highway network bubble to pop? Sadly the answer is "yes", because of enrichment from two possible sources: additional construction and the transition from internal combustion engines to electric motors expected to move 21st century cars and trucks. Both of these alter the Interstate's carrying capacity in negative ways.

---

[4] http://en.wikipedia.org/wiki/Interstate_Highway_System

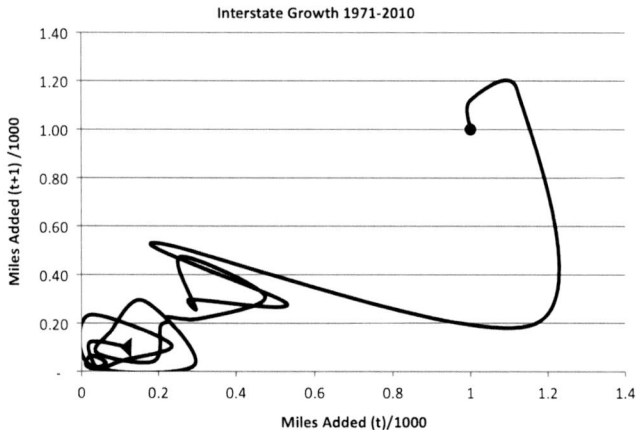

Fig. 5.7 State-space diagram of new construction of roads in the Interstate Highway network shows the system spiraling down and nearing a fixed point

The original Interstate freeway plan called for 41,000 miles. Today it is nearly 50,000 miles and growing. Proposals are pending for Interstates 3, 9, 11, 14, 22, 41, 67, 92, and 98. But, according to Fig. 5.7, the miles of new pavement added each year are dwindling. Figure 5.8 shows the relationship between funding and extensions of freeway: the state-space diagram shows they are approaching a fixed point.

The state-space diagram of Fig. 5.7 shows Interstate Highway miles decreasing from a peak construction rate in 1981 to a fixed point near zero in 2010. The squiggly circles also indicate a turbulent history of funding and construction. There have been starts and stops, as waves of construction came and went throughout the 1980s and 1990s. To make things even less stable, recent acts of Congress have begun to shift Highway Trust Funds from roads to other forms of transportation like buses and rail.

The state-space diagram of Fig. 5.8 shows the predator-prey relationship between construction and funding. Throughout the 1960s and 1970s miles were vigorously added, but then extensions to the nation's main arteries were restricted. Congress began raiding the highway funds to pay for busses and commuter trains. Growth leveled off, and then during the first decade of the 21st century, freeway construction stalled, just when it turned 50 and began to decay. It appears that both construction and funding have reached a fixed point. Funding and construction are stuck at this fixed point, because most of the funds are now being spent maintaining the "overgrazed" system.

What happens to the system when maintenance overwhelms construction or electric cars deplete funding? A projected state-space diagram in Fig. 5.9 shows what can happen when enrichment due to increased construction of "more miles"

5.8 The Road to Oblivion

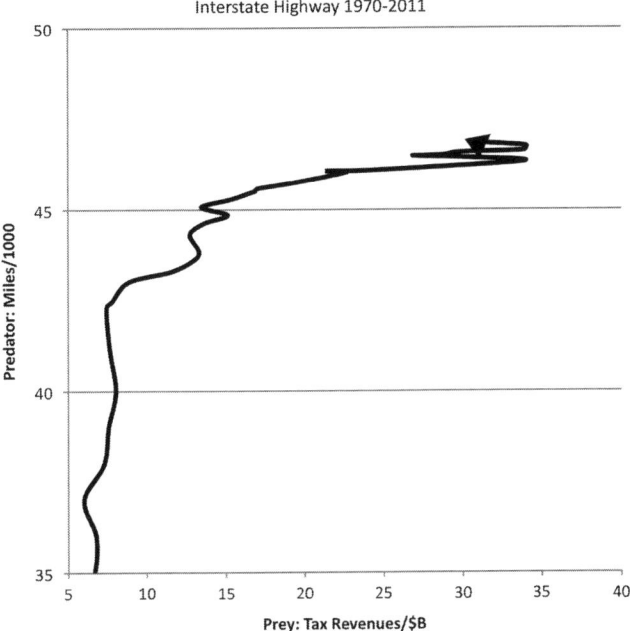

Fig. 5.8 State-space diagram of the predator-prey relationship between construction of roads and funding in the Interstate Highway network shows the system approaching a fixed point

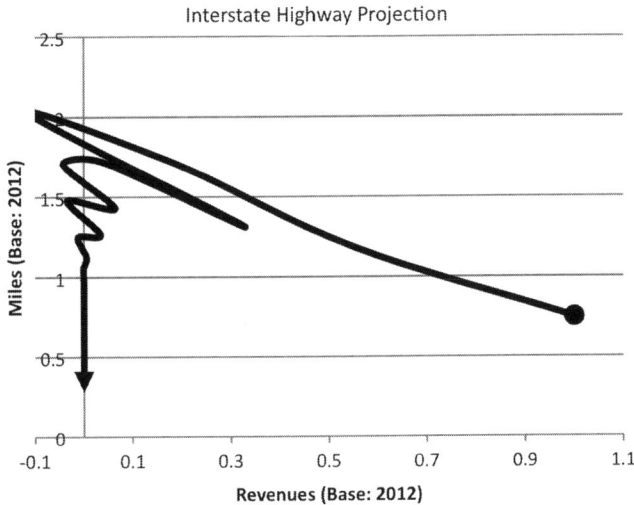

Fig. 5.9 State-space diagram of a possible Interstate highway extinction due to replacement of gasoline fueled vehicles with electric vehicles and deterioration of existing roads

outstrips funding. The state-space diagram assumes construction funding diminishes because of the increasing burden of maintenance or because gasoline-powered cars are replaced by non-taxable electric cars. In either case, the "highway bubble" pops due to the induced instability.

Figure 5.9 is hypothetical, of course, but it is based on mild assumptions. First, funds derived from taxing gasoline could decline by 5 % per year. This is a reasonable assumption if electric vehicles replace gasoline vehicles in twenty years and currently levied gasoline taxes are not replaced by some other tax. Second, it assumes that maintenance begins to deplete roads at a rate of 10 % per year. While this seems high, it may not be: most of the Interstate Highway system is already near the end of its 50-year lifecycle.

Figure 5.9 also illustrates how instability in both revenues and length of highway perturbs the highway ecosystem in nonlinear ways as replacement and decay of existing roads advances. The oscillations in road length as funding dwindles and approaches zero may become even worse if the assumed rates of vehicle replacement or road decay accelerates because government subsidizes electric vehicles or if the US experiences a reduction in gasoline consumption. Once again, this underscores the fragility of a relatively simple, but complex, system.

Something as simple as a highway network may be subject to the paradox of enrichment and tragedy of the commons.

5.9 A Moral Hazard

On March 11, 2011, one of the 15 largest nuclear power plants in the world was severely damaged and closed down by a 9.0 earthquake and subsequent tsunami. The long-term effects of nuclear radiation leaks are unknown, but at the time I write this, a 20 km square area of Japan is uninhabitable. A month later Japan's National Policy Agency counted 11,620 dead and 16,464 missing persons. The official number of people evacuated exceeded 114,000 as of December 2011. Estimates of the damage range from $250 to $650 billion. The reactors themselves were worth approximately $50 billion.

Is this extreme event the end of nuclear power? The answer for the United States is 'no', because of the nuclear power commons established in 1959. While a Fukushima-magnitude disaster is highly unlikely to happen again, it is almost certain that a Three-Mile Island-magnitude nuclear accident will happen sometime in the future. But the *Price-Anderson Nuclear Industries Indemnity Act of 1957* protects investors and owner/operators from liability in excess of $111.9 million—a fraction of the liability. There are 104 USA Power Reactors easily worth $20 billion each, but the US government has limited owner/operator risk to $111.9 million, or one-half of one percent of a typical power plant's replacement cost.

The US nuclear power ecosystem is an industrial commons. It may also suffer from tragedy of commons complexity. Here is why. Power reactor licensees are required to insure the first $375 million in loss per reactor due to a failure and pay up to $111.9 million per reactor in the event of an accident on claims that exceed the $375 million insurance policy. Actual payments are capped at $17.5 million per year, until $111.9 million or the amount of the liability is reached. This fund is not paid into unless an accident at one of the 104 plants occurs.

But here is the part that makes this a potential tragedy: all owner/operators are required to pay if any one of them suffers a catastrophic loss. In other words, losses are shared, but profits are not. However, owner/operator liability is limited to $111.9 million, or a total of $12.2 billion ($111.9 m × 104 reactors), even if the total cost is $250–$650 billion as it was for Japan. Who makes up the difference?

Wikipedia defines *moral hazard* as, "a situation where there is a tendency to take undue risks because the costs are not borne by the party taking the risk". The US nuclear power industry fits this definition. In fact, the *Price-Anderson Nuclear Industries Indemnity Act* was written to motivate the power industry to build nuclear power plants in the face of high risk. This regulation was perhaps the original "too big to fail" moral hazard put in place by the US government to persuade the power industry to "take undue risks" on purpose.

The Fukushima dai-ichi meltdown may end nuclear power in Japan, but not the US. The industry enjoys protection from a kind of tragedy of the commons, because risk has been shifted onto taxpayers. Could this change? If a nuclear nightmare of the same magnitude as Fukushima dai-ichi happens in the US, the bubble could burst. But until then, expect US nuclear power reactors to stay in business.

The Price-Anderson Nuclear Industries Indemnity Act was renewed in 2005 for another 20 years.

5.10 The Dutch Disease

Economic systems are almost as complex as biological systems. We tend to under estimate economic complexity because the "law of supply and demand" is so simple. In fact, economic complexity requires careful analysis before making policy that may cross a tragedy of the commons tipping point, increase self-organized criticality, or trigger enrichment. The famous *Dutch Disease* is a case in point.

An article appearing in *The Economist* in 1977 coined the term "Dutch Disease" to describe the decline of the manufacturing sector in the Netherlands after the discovery of a large natural gas field in 1959.[5] Discovery of a valuable resource like natural gas should be cause for celebration, but as it turned out, the discovery

---

[5] The Dutch Disease (1977).

triggered a form of enrichment and distorted the Dutch economy by exceeding its carrying capacity. But it did so, indirectly, through a complex interaction.

Here is what happened. The natural gas discovery sent prices of everything through the roof. The natural gas sector thrived, but all other sectors suffered from the tragedy of the commons by raising costs across the entire economy. The corresponding cost-of-living boom drove up labor rates and commodity prices, which in turn, eradicated prosperity in the broader economy. By enriching one sector, the discovery crashed other sectors.

W. Max Corden and J. Peter Neary developed a general economic theory in 1982 to explain bubbles like the Dutch Disease.[6] In their model, there are two sectors—a *booming sector* and a *lagging sector*. The booming sector is usually based on natural resources like natural gas, oil, gold, copper, or agriculture. The lagging sector generally refers to manufacturing. In terms of the predator-prey paradigm, two predators (booming and lagging sectors) compete for one prey (labor).

Here is how enrichment leads to the rise and fall of an economy through an imbalance between booming and lagging sectors. A surge in the booming sector ultimately increases the demand for labor, which increases salaries in the sector, which then shifts production away from the lagging sector. For example, during a gold rush the mining sector will draw labor away from the manufacturing sector. To compound the problem the booming sector drives up the cost of labor, which further punishes the lagging sector. Wages in the lagging sector are set internationally, so this sector is caught in a two-pronged trap—low wages and high labor shortages. So if the booming sector is gold mining, the automobile sector cannot compete for labor because the price of automobiles is set by international competition and not the price of gold. Thus, the automotive sector languishes and perhaps goes out of business.

Dutch Disease also contains elements of *Gause's competitive exclusion principle*. As you recall, this principle says only one dominant species can emerge from an ecosystem starting with a field of multiple competitors. In this case the booming and lagging sectors compete for labor. Eventually, the booming sector wins because of its advantage—it pays higher wages.

Whether caused by enrichment or temporary booms, bubbles eventually pop and send waves of consequence in all directions. Enrichment can capsize an economy by exceeding its carrying capacity, and enrichments like the Dutch Disease can capsize an economy by squeezing out less-competitive sectors. They may even spread nonlinear effects to other economies throughout the globe. Does a market crash in one country also capsize the economies of other countries? Do financial contagions begin with bubbles and end with global meltdown? That is the question I answer in the next chapter.

---

[6] Corden (1984).

References

Corden W. M., (1984). "Boom Sector and Dutch Disease Economics: Survey and Consolidation". Oxford Economic Papers 36: 362.

Hardin, Garrett J. (1968), The Tragedy of the Commons. Science 162, 1243–1248.

Lahart, Justin (2007), "In Time of Tumult, Obscure Economist Gains Currency", The Wall Street Journal, (2007-08-18), http://online.wsj.com/public/article/SB118736585456901047.html

Minsky, Hyman P. (1992), The Financial Instability Hypothesis (May 1992). The Jerome Levy Economics Institute Working Paper No. 74. Available at SSRN: http://ssrn.com/abstract=161024 or http://dx.doi.org/10.2139/ssrn.161024.

"The Dutch Disease" (November 26, 1977). The Economist, pp. 82–83.

Shocks 6

Abstract

If national economies can be rocked by the Paradox of Enrichment, what happens to other countries inextricably connected to failed economies? Can a crash in one part of the world spread to other parts? Do shocks propagate like a disease? The 21st century is driven by David Ricardo's theory of comparative advantage. It is the motivation for globalization and the connected world. One of its most severe consequences is the "contagion effect", which claims to unsettle every nation in the world, when one nation suffers a setback. Globalization yields a World Trade Web of interconnected countries. It also establishes a contagion vector along which financial (and social) shocks travel. But, as it turns out, the impact of a shock in one country depends on the structure of the World Trade Web. Only the most-highly connected trading countries can survive global shocks that occur with long-tailed probability.

6.1 The Richest Economist in History

He was the third of 17 children in an age of turbulence and war. His family disowned him when he eloped with his Quaker lover. His father cut him off from the family fortune and his mother never spoke to him again. His life was rather dull and uneventful until he read Adam Smith's *Wealth of Nations* while vacationing in Bath, England. At 37 years of age, his fame took off after turning his attentions to the emerging discipline we now know of as *economics*.[1]

The richest economist in history made a million pounds in one day betting on war bonds. During his most productive decade he profited from profligate government spending—financial exuberance that sent the country into economic

---

[1] Economics is the study of production, consumption, and transfer of wealth.

chaos. One might say he made his living by betting on spikes in government borrowing and the inevitable shocks that follow excessive government spending. If this sounds like a modern tale, your are correct. But it happened over 200 years ago.

David Ricardo (1772–1823) was an arbitrage king—a trader that buys low and sells high. Arbitrage is the art of profiting from imbalances that inevitably occur between buyers and sellers. This form of "day trading" is especially profitable during choppy economic times. And Ricardo lived during very choppy times. His fortune soared during the Napoleonic War years (1811–1815) reaching a peak when he bet against Napoleon at Waterloo. Ricardo's hedge paid off in seven digits when financiers learned of Admiral Nelson's victory. At the time of his death in 1823 Ricardo's fortune exceeded that of John Maynard Keynes—the second richest economist in history.

Today's exuberant bond market reminds us of Ricardo's time. Vast sums of money change hands as governments borrow trillions of dollars to finance wars, social programs, and bailouts. England's national debt was 250 % of its GDP in 1815 when Ricardo cashed in his bonds. The economy was in such bad shape that the government temporarily suspended taxes. [In 2014 the US debt ratio exceeded 100 % of GDP, depending on how debt is measured].

Sound familiar? Today's bond kings take advantage of the "spread" created by low-interest government money and ever-increasing public debt. Central banks enrich the money supply, which create bubbles destined to burst. Markets rush from commodities, to gold, to equities, and back again to currencies. Unemployment yoyo's as bubbles rotate from sector to sector. As a result, the economy experiences a series of shocks—punctuations that upset markets and rattle consumers—not to mention voters.

The question posed in this chapter is simply this, "do economic shocks matter?" The previous chapter described the underlying causes of enrichment. This chapter explores its consequences. Can a nation such as the United States go on borrowing and spending? What happens to the world economy if the US dollar, which accounts for the majority of all currency in circulation in the world—suddenly plummets? Do economic shocks cause great depressions and global economic collapses? Or do they dissipate as quickly as they appear?

6.2 Comparative Advantage

Ricardo gave the world two major economic gifts: the *theory of comparative advantage*, and the *theory of diminishing returns*, which he saw as the manifestation of saturated demand. Comparative advantage was little more than a theoretical construct in 1810, but it is the driver of economic power in the 21st century, because it is the source of wealth of modern nations. But it is sometimes throttled back by Ricardo's second theory—consumers soon become satiated and demand for goods plunges, leading to chaotic *paradox of enrichment*. Comparative

6.2 Comparative Advantage

Table 6.1 Exchange of iPads and Shirts benefit both the US and China

	iPad	Shirts	Totals
(a) *Protectionist*			
China	1,000	500	1,500
US	500	800	1,300
(b) *Free trade*			
China	500	500	1,000
US	500	500	1,000

advantage stimulates wealth while diminishing returns stunts it. It is the tension between these two forces that create shocks to the global financial system.

Comparative advantage is simple but powerful. During Ricardo's day the British Empire ruled the world's economy through a series of protectionist measures such as the infamous Corn Law and through control of the global supply chain. The Corn Law protected English farmers by restricting imports of cheap wheat from the colonies. [The Corn Laws actually did more harm than good and were eventually repealed]. The British navy ruled over the high seas and guaranteed the security of trade routes leading to the British shores, thus securing Britain's supply chain. But British politicians failed to understand Ricardo's economic theories and like all fearful people, reacted with restrictive policies that made things worse.

Ricardo advocated what we now call *free trade*, arguing that trade benefits both buying and selling countries, because of each country's comparative advantage. Table 6.1 illustrates how it works. Consider two scenarios. In the first scenario—summarized in Table 6.1a—we assume the equivalent of iPads and Shirts are designed, developed, and sold in China for $1,000 and $500, respectively. It would cost Chinese consumers $1,500 to acquire both if domestically produced. Similarly, assuming iPads and Shirts cost $500 and $800, respectively, if produced within the US, US consumers would pay a total of $1,300 for both. But both nations can benefit by trading, rather than producing completely within its borders.

Now consider Table 6.1b, which assumes China and the US engage in international trade. If Chinese consumers purchase US-made iPads they reduce their cost from $1,000 to $500. Similarly, if US consumers purchase Chinese-made Shirts they reduce their cost from $800 to $500. Both economies benefit—the Chinese save $500 and the American's save $300. Each country leverages its advantage. China is rewarded for buying iPads from the US and the US is rewarded for buying Shirts from China. The saved Yuan and dollars can be redirected to more profitable and beneficial products in each country.

Comparative advantage is the engine of economic prosperity for trading countries. The more they trade, the more they prosper. How so?

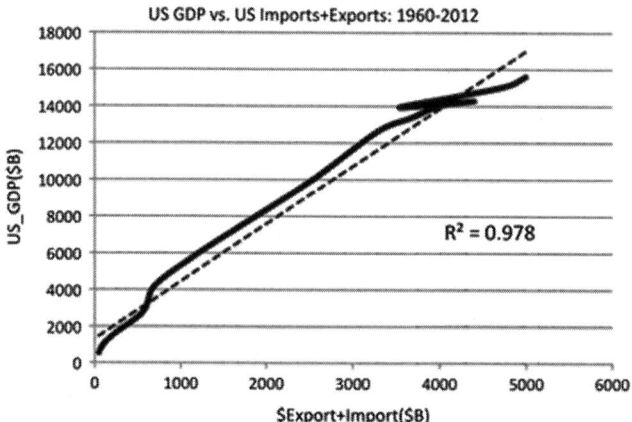

Fig. 6.1 US GDP is strongly correlated with trade. Correlation diminishes when enrichment introduces nonlinearity into the growth of an economy

6.3 Extreme Economics

The benefits of comparative advantage and free trade explains why there is such a strong correlation between trade and GDP, see Fig. 6.1. The correlation is strong—more trade equals more GDP, especially when comparative advantage is fully exploited. But the correlation is not linear and not without punctuations. For example, the correlation between the European Union country's GDP and EU exports shown in Fig. 6.2 contains a linear part, but also a strong nonlinear part.

It would seem that EU economics are a bit extreme. Why?

6.4 GDP: A Measure of Fitness

Gross Domestic Product (GDP) is often derided as an inadequate measure of wealth and prosperity, but I will use it anyway, because it is an adequate measure of economic fitness. First developed by Simon Kuznet in 1934 and then widely adopted in 1944 following the *Bretton Woods Accord*, GDP has become the main means for quantifying a country's economic wellbeing. In simple terms, GDP is the sum of all monetary transactions that go into calculating "wealth", as follows:

$$\text{GDP} = \text{private consumption} + \text{investment} + \text{government spending} + \text{trade}$$

Private consumption, investment, and government spending are byproducts of a prosperous economy. They all increase when times are good and they (mostly) decrease when times are bad. Government policies regarding research, education, and waging of wars play a major role in shaping GDP, but note that trade has been

6.4 GDP: A Measure of Fitness

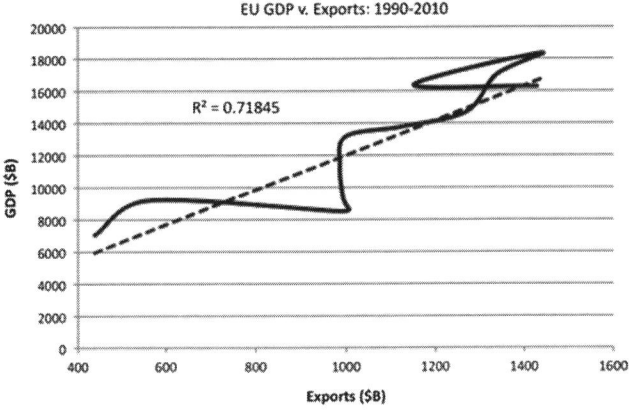

Fig. 6.2 Correlation between EU GDP and EU exports contains a linear and nonlinear part. The first nonlinear episode occurred in 2001 and the second in 2008–2009

the most influential factor during recent times, see Fig. 6.1. GDP is correlated with import and export activity—especially with regard to exports. Almost all of the fluctuations in GDP can be explained by "anomalies" in export activity.

In simplistic terms, trade establishes a virtuous cycle. Raw materials such as oil and metals are imported into manufacturing nations that add value by producing finished goods that are exported back to other nations according to their comparative advantage. The raw-material country benefits and so does the manufacturing country. Comparative advantage lowers the cost of raw materials and also lowers the cost of finished products so both nations benefit. The connection is not exact, however as suggested by Figs. 6.1 and 6.2.

Figure 6.1 raises two important questions. First, why is there such a strong linear correlation between GDP and trade, and second, what causes the occasional departure from a straight line? I hope to convince you that the linear correlation is "normal" and to be expected because trade–the comparative advantage of exports especially—drive the economy. The linear correlation evident in Fig. 6.1 verifies Ricardo's theory of *comparative advantage*. But what about the deviations from linearity also evident in Fig. 6.2? Ostensibly they were caused by the dot com crash of 2001 and the 2008 financial meltdown. How powerful were these shocks? Can future shocks ruin the global economy?

Economic theories adequately explain the economy when the complex system we call "economics" is working smoothly. But when a shock such as an enrichment or depression comes along, most economic theories no longer work. Like a python devouring a pig, the economic framework of an entire nation—and sometimes the entire globe—becomes stretched and distorted. Traditional economics is unable to explain these extreme distortions. Instead, we have to turn to non-classical complexity theory. Classical economics has to be extended into the quirky realm of *chaos theory*.

What causes the punctuations in Figs. 6.1 and 6.2?

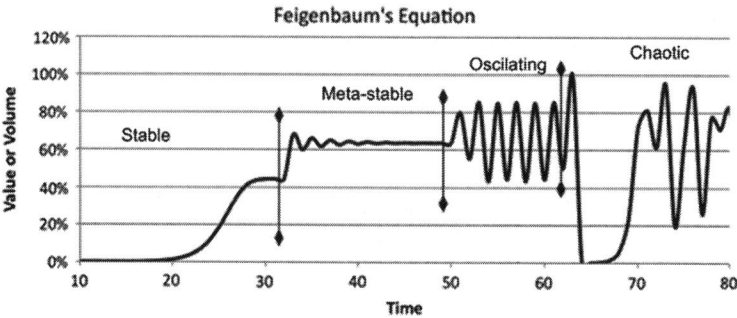

Fig. 6.3 The Feigenbaum equation illustrates various regimes of behavior: from stable to meta-stable, meta-stable to oscillating, and oscillating to chaotic or "out of control"

6.5 Chaos Theory

Chaos theory actually has little to do with chaos as defined by Merriam-Webster.[2] Chaos theory got its name because some systems are so unstable their behavior mimics randomness. But mathematical chaos theory is more precisely the study of nonlinear system dynamics. Nonlinearity comes in several flavors—meta-stable, oscillating, and chaotic.

Figure 6.3 illustrates the difference between stable, meta-stable, oscillating, and chaotic systems. Stable systems change smoothly as illustrated by the first section of Fig. 6.3. Meta-stability introduces wiggles as shown in the second section, and rhythmic oscillations swing up and down equally as shown in the third section. A rhythmic pattern is discernable in all but the chaotic portion of Fig. 6.3. The chaotic portion, however, gyrates from one extreme to another, unexpectedly, without repetition or discernable patterns. Chaotic systems appear to behave randomly, but in fact they are highly predictable—if we only understood the nature of nonlinearity.

Belgian mathematician Pierre François Verhulst (1804–1849) studied the earliest known equation capable of chaotic behavior in 1838. He called it a *logistical growth function* in a paper published in 1845 in which he modeled growth of animal populations when confronted by limitations such as food, water, and space. Verhulst realized that in every complex system there is a period of rapid growth, followed by a peak, followed by a decline, and then possible extinction. His logistical growth curve is an S-shaped curve as illustrated in the stable section of Fig. 6.3. Populations rapidly grow and ultimately hit resource limits. These limits dampen further expansion and either cause leveling off or precipitous decline [Shades of Ricardo's diminishing returns!].

---

[2] Chaos is the state of things, in which chance reigns supreme.

Verhulst's model has been rediscovered by dozens of scholars in dozens of disciplines over the past 150 years. For example, American bio-gerontologist and prolific science writer Raymond Pearl (1879–1940) and John Hopkins University colleague Lowell Reed (1886–1966) promoted Verhulst's logistical growth equation in the 1920's to explain various theories in biology. A similar curve was discovered slightly before Verhulst's time in 1825 by a self-educated British mathematician and actuary, named Benjamin Gompertz (1779–1865). The *Gompertz law of mortality* is still used to day. The *Gompertz curve* has also been used to explain the adoption of mobile phones, where costs are initially high followed by a period of rapid growth, followed by a slowing of sales as market saturation is reached. [Ricardo's diminishing returns, again].

The growth-peak-death idea spread to many other fields. Everett M. Rogers (1931–2004), sociologist, writer, and teacher popularized the equation once again in a 1962 book, *Diffusion of Innovations*. Roger's name is familiar to many Silicon Valley entrepreneurs because he coined catchy phrases like *early adopter*, and *first-mover*. Frank Bass (1926–2006) also applied the equation to diffusion of products such as the personal computer and *iPods* that "take off", initially reach a peak, and then decline to zero over a period of time. That is, they follow a *lifecycle* that obeys the stable S-shaped curve of Fig. 6.3 (adoption), and its rate of change (sales).[3]

Finally, the logistical growth curve has been applied to resource-limited industries such as energy exploration and exploitation. The *Hubbert curve* describes so-called *peak oil*—the rise and fall of oil reserves around the globe. In 1956, Marion King Hubbert (1903–1989), a geologist working for Shell Oil research in Houston, used a modified version of the logistics growth curve to predict that US oil production would peak in 1970. Hubbert's peak oil equation describes the rise and fall of oil reserves for a given region.

What do all of these models have in common? They describe systemic shocks—a sharp increase and fall in population, oil resources, or GDP. The rapid rise is easy to understand, but the correspondingly rapid fall is more difficult to explain. Must a rise always be followed by a fall? Why are some shocks "normal", while others are extreme and chaotic? What property of a system makes it more likely to collapse?

The logistical growth functions studied by Verhulst and others turn out to be inadequate for understanding the shocks that come and go in the 21st century. The missing ingredient was not identified until the 1970s when Rosenzweig (1941–) discovered the *paradox of enrichment*, and Mitchell Feigenbaum (1944–) introduced us to *chaos theory*. Feigenbaum's modification to the logistical growth function explains the wrinkles observed in Figs. 6.1 and 6.2.

Instability turns a smooth and well-behaved logistical growth curve into a jagged and punctuated Feigenbaum curve.

---

[3] An S-curve is obtained by integrating the rise-and-fall equation over time.

6.6 Master of Chaos

Winifred Burkle—heroine of the 2011 TV series *Angel*—called her stuffed bunny the *master of chaos* in an episode titled, "A Hole in the World". She was talking about a real-world person, Mitchell Jay Feigenbaum (1944–), famed mathematical physicist, Toyota Professor at Rockefeller University of New York and recipient of the MacArthur Fellowship (1983) and Wolf Prize in Physics (1986). Feigenbaum began experimenting with the famous logistical growth equation using an HP-65 calculator in 1975. His findings were published in a now-famous paper, "Quantitative Universality for a Class of Nonlinear Transformations."[4] This is a must-read for every serious systems-thinking person.

Feigenbaum ignored the limits put on the famous logistical growth equation by its pioneers. The traditional view of growth was that a population or resource could never exceed 100 % of its capacity. But Feigenbaum removed that constraint and made a leap in thinking that will help us understand shocks. What if a population or some other resource-limited phenomenon exceeded its limits for a short period of time? That is, what happens when a system goes out of bounds? Not every system "blows up" when this happens, but the logistical growth function does under certain conditions. It becomes chaotic.

Feigenbaum found a *tipping point*, 3.57..., in Verhulst's logistical growth function which led him to the famous *Feigenbaum number*, 4.669201609..., representing the point at which the general class of logistics functions *bifurcate*, or split into different regions—from stable to chaotic. The bifurcation regions are illustrated in Fig. 6.3. As long as rate of growth and *carrying capacity* of a system remain within certain bounds, the system responds smoothly along an S-shaped curve. But when the growth rate, carrying capacity, or both exceed certain bounds, the system responds erratically. At first it begins to wobble or oscillate. Then it continuously oscillates, and finally it "blows up" as shown in Fig. 6.3.

The pioneers of logistical growth curves had previously overlooked the subtlety of nonlinearity lurking in the 150-year old equation. Not so Mitchell Feigenbaum, who promptly called his parents to announce he had "discovered something truly remarkable." In fact he had, which brings my story back to extreme economics. Feigenbaum's truly remarkable discovery can be used to explain the effect of shocks on the wealth of nations.

Fundamentally, the nonlinear distortions observed in Figs. 6.1 and 6.2 are caused by enrichment. Enrichment temporarily exceeds the bounds of logistical growth and introduces nonlinearities, which manifest as wrinkles and oscillations in GDP. GDP goes up and down with exports, but when government or businesses temporarily introduce an "out of bounds" period of enrichment, Feigenbaum's chaos steps in. So GDP is a combination of stable and unstable logistical growth, see Figs. 6.4 and 6.5.

---

[4] Feigenbaum and Mitchell (1978).

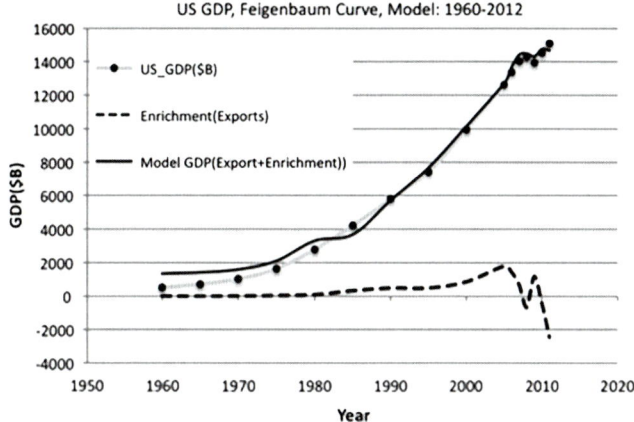

Fig. 6.4 US GDP modeled as a combination of straight-line growth and enrichment. Governments attempt to implement various forms of enrichment, which result in chaotic adaptation. The *dashed line* is a Feigenbaum curve representing the effects of (unstable) enrichment

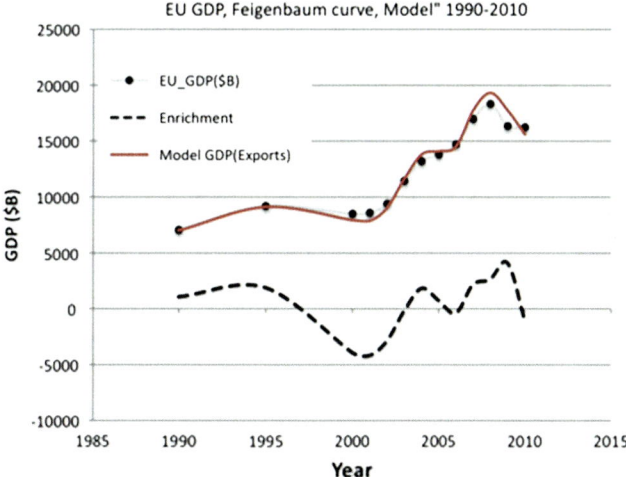

Fig. 6.5 EU GDP modeled as a combination of straight-line growth and enrichment. The dashed line is a Feigenbaum curve representing the effects of (unstable) enrichment. Compared with the US model in Fig. 6.4, EU governmental attempts to "fix the economy" by various forms of enrichment have led to more destabilization than similar attempts by the US government. Why are some countries more immune to shocks than others?

Extreme economics is an exercise in chaotic instability in at least one sector of the economy. Whenever a sector of the economy "goes ballistic" or rises exponentially, the inevitable sets in and the balloon bursts. This nonlinearity can only be explained by substituting extreme economics for classical economic theory.

6.7 Tequilas and the Asian Flu

The *Tequila crisis*—a shocking devaluation of the Mexican peso in December 1994—produced a lot of bad press but failed to shock the rest of the world. It was caused by enrichment—a decade of hyperinflation, heavy government debt, ultra low oil prices, and a shortage of cash. It took Argentina down with it, but the rest of the world was virtually unscathed. Why didn't the Tequila crisis spread throughout the rest of the world?

The 1997 *Asian Flu* gripped much of Asia following a financial crisis in Thailand. High foreign debt crashed the Thai baht and bankrupted the government. The "Asian economic miracle" of the previous decade was fuelled by classical enrichment gone to extremes. The Southeast Asian banks charged high interest rates, which attracted large investments, which in turn supercharged their economies. During the ramp up to the crash countries like Thailand, Malaysia, Indochina, Singapore, and South Korea grew at rates of 8–12 % annually. But the Asian Flu soon flamed out before contaminating the entire global financial system. Why did the Asian Flu make a bigger impact than the Tequila crisis?

The Russian Virus of 1998 also flared up and died out without capsizing the world's financial system. In fact, all of the shocks described here had disastrous effects on certain countries, but not others. Argentina, Venezuela, Thailand, and Mexico were heavily impacted, but the US, China, India, Germany and UK economies were not. What determines the impact of financial shocks in one part of the world on the rest of the world? What stops a contagion?

6.8 World Trade Web

The extreme economics model of GDP growth contains two parts—a linear part that conforms to Ricardo's classical theory of comparative advantage, and a nonlinear part driven by non-classical, nonlinear chaos. GDP growth is modeled by a straight-line derived from the linear growth of GDP (Figs. 6.1 and 6.2) and the nonlinear Feigenbaum equation as shown in Figs. 6.4 and 6.5. Note the very close match with observed values.

Generally, GDP increases with increases in trade. Occasionally, the economy of one or more nations is disrupted by either internal or external events, and trade spikes—either up or down. When a system undergoes a shock of significant size growth becomes nonlinear and Feigenbaum's chaos sets in. Chaotic instability may persist or not, depending on the actions of government, productivity of the people, and the connectivity of nations as defined by their trade relationships.

Simple models of national wealth may explain one country's GDP in isolation, but what does it say about the global flow of goods and services we call international trade? What impact does one country have on its trading partners? To answer this system-level question, we need to construct a WTW—*world trade web*. The WTW is a network, as shown in Fig. 6.6, where nodes represent

6.8 World Trade Web

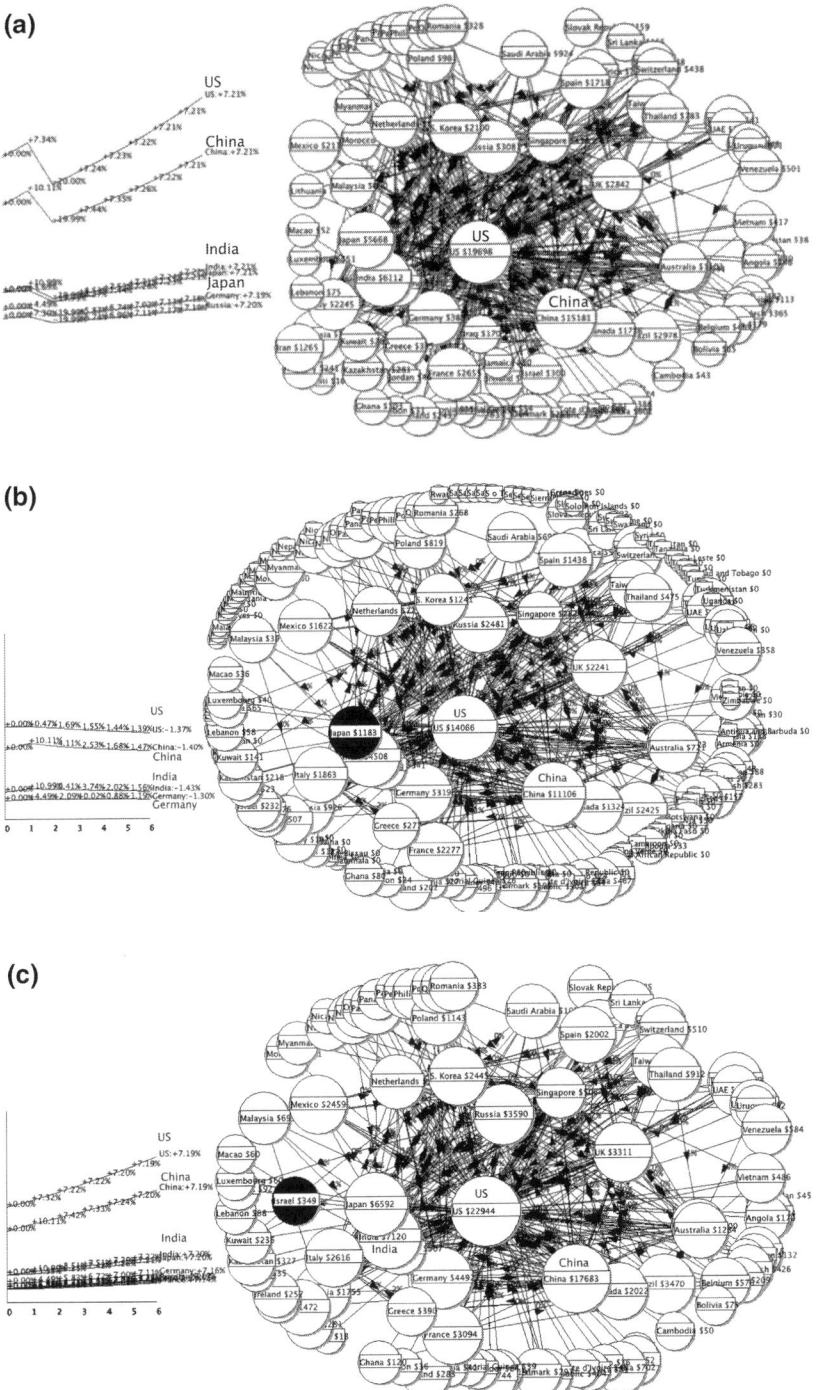

Fig. 6.6 The WTW—World Trade Web is a network of trading nations and their import-export links. Simulation **a** shows the effect of a 20 % shock to the USA economy; **b** shows the effect of a stagnant Japan; and **c** shows the effect of a 20 % shock to Israel's economy

countries and links represent import/export relations between pairs of countries. The size of each node in Fig. 6.6 is proportional to its GDP.

Angeles Serrano and Marian Boguna of the Universitat de Barcelona, Barcelona, Spain showed that the WTW is a complex network[5]: it is *scale-free*, wired together like a *small world* (small diameter), and contains clusters of nodes representing regional trading partners. Fortunately, it is also rather resilient against shocks in GDP, as I will demonstrate with a simulation.

Figure 6.6 is a network of trade connections where the GDP of a country depends on how much it exports to its trading partners and how much its economy is enriched. In simple terms, changes in GDP are the sum of growth, exports, and enrichment:

GDP $\sim$ Growth + Export revenue + Enrichment
Growth $\sim$ organic growth rate
Export revenue $\sim$ sales of exports
Enrichment $\sim$ carrying capacity

Figure 6.6a shows the result of a 20 % drop in the industrial output of the USA, similar to the 2008 financial meltdown that bottomed out in March 2009. The shock spread to China and other major trading nations, but lasted for only a short time. On the contrary, a stagnant Japan, as shown in Fig. 6.6b has a more persistent moderating affect on the major economies of the world. But if the economy of a small trading country like Israel slows or drops precipitously, the economies of the world barely notice.

Simulation of the WTW uncovers an obvious truth: big trading countries can damage others, but small trading countries have little effect on the world's economies. Furthermore, big economies (and traders) spread contagions while small economies (and traders) do not. "Big" is defined here by number of trade links or node degree.

Simulation of the WTW shown in Fig. 6.6 suggests that the wealth of a nation is directly correlated with amount of trade—both imports and exports. The so-called "rich club nations" conduct more trade than the so-called "emerging nations". Trade intensity is equal to the number of trade links connecting a country to other countries. Thus, we can rank countries according to their connectivity or network *degree*, because more links equals more trade. High degree countries are big traders.

Figure 6.7 summarizes the results of a number of simulations performed on the WTW of Fig. 6.6. Each scenario studied spans the period 2012–2020 and assumes growth rates that existed in 2011. Shocks are defined as a 20 % drop in GDP in year 2 of the simulation. The results of these simulations are summarized, below:

Scenario #1: A no-growth policy in Europe reduces GDP of the selected countries in Fig. 6.6 by 5–10 %. The sovereign debt crisis that plagued Europe following the 2008/9 shock has little effect on global "financial contagion". Why?

---

[5] Serrano et al. (2003).

6.8 World Trade Web

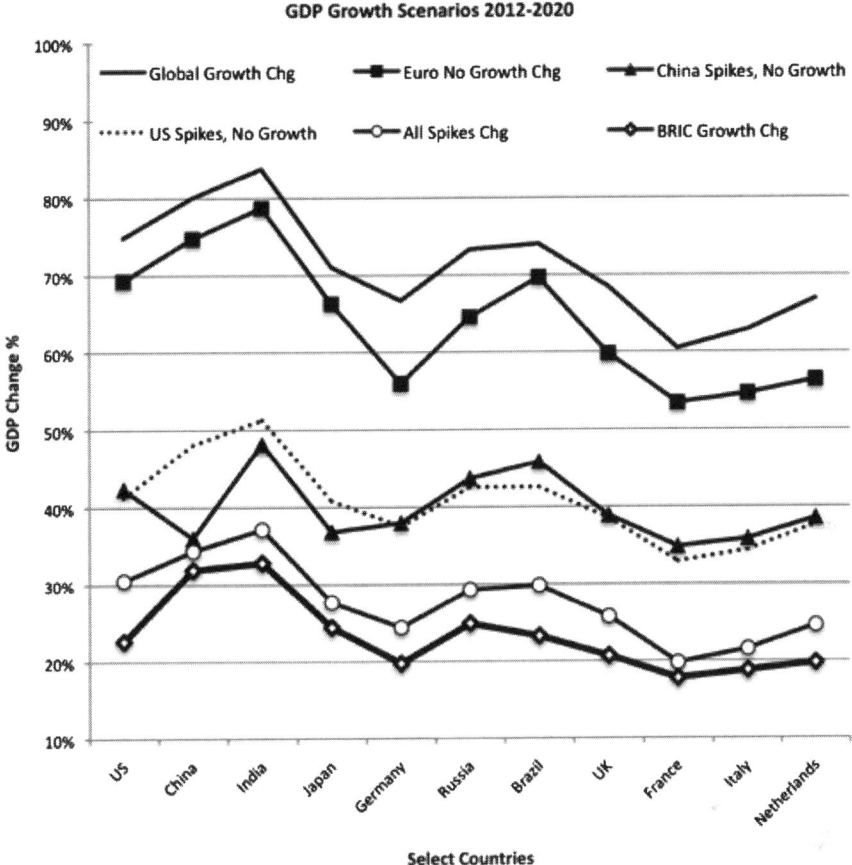

Fig. 6.7 Scenario results of simulation experiments performed on the WTW of Fig. 6.6a China and India show the most growth in GDP regardless of the scenario—Europe shows the lowest growth

Scenario #2: A 20 % shock (drop) in US or China GDP produces a 25–30 % drop in GDP by 2020 compared to the 2020 GDP anticipated without a spike in 2012. China and the US shocks have a major impact on the global financial network. Why?

Scenario #3: If all countries experience a 20 % drop in GDP in year 2 of the simulation a 40–50 % drop, should be expected, globally. This is no surprise, but provides a benchmark for comparisons.

All Scenarios: BRICS (Brazil, Russia, India, and China) survive shocks 5–15 % better than Europe, mainly due to their rapid growth. No surprise here, either.

Table 6.2 Most-vulnerable and most-influential trading nations in the WTW are determined by the number of trade links of a country

Most vulnerable	Trade links	Most influential	Trade links
Egypt	4	US	112
Iran	8	China	29
Taiwan	8	India	25
S. Africa	8	Japan	33
Saudi Arabia	8	Germany	30
Thailand	9	Russia	27
Mexico	10	Brazil	11
Poland	11	UK	36
Brazil	11	France	19
Spain	14	Italy	20

These scenarios show that some financial contagions are worse than others. What determines the impact of a "financial contagion?" Stefano Schiavo and colleagues at the University of Trento, Trento, Italy, claim, "International connectedness [alone] is <u>not</u> a relevant predictor of crisis intensity".[6] In fact, they concluded the reverse, "adverse shocks dissipate quicker" for countries with more trading partners. Rather than spreading financial contagion faster, a WTW country with many trading links tends to dissipate financial contagion. "We show that higher interconnectedness reduces the severity of the crisis, as it allows adverse shocks to dissipate quicker. However, the systemic risk hypothesis cannot be completely dismissed and being central in the network, if the node is not a member of a rich club, puts the country in an adverse and risky position in times of crises. Finally, we find strong evidence of nonlinear effects."

In other words, big trading countries stabilize shocks created by minor trading countries.

6.9 The Raging Contagion Experiment

To confirm the expert's claims, I performed another simulation on the WTW as follows. The computer simulator repeatedly selects an epicenter country at random and marks it as "contaminated". Then it spreads the contamination to its trading partners with probability 1/d, where d is the degree (connectivity) of the partner. The simulator repeats this thousands of times, tallying the number of times trading partners are contaminated. The tallies are proportional to the probability that country A will be rocked by a shock in country B. These tallies are a measure of vulnerability to financial contagion. The results are shown in Table 6.2.

---

[6] Chinazzi et al. (2012).

Now you know the answer to the "financial contagion" question. The most vulnerable countries in Table 6.2 are those with the fewest trading partners. The most influential countries have the most trading partners. In other words, highly connected countries are super-spreaders of financial contagion, but also resilient against incoming contagions. Conversely, sparsely connected countries have little impact on other countries, but are more vulnerable to incoming contagions. Strength and power are both vested in the number of trade relations maintained by a country.

This experiment argues for governments that advocate a policy of free trade among nations, because everyone benefits. It also warns against economic warfare, because the collapse on one economy can boomerang.

(a) The WTW shows the effect of a 20 % drop in US GDP on the global trading network. A plot of GDP versus time is shown on the left, and the WTW, with 178 countries and 372 trade links, on the right.
(b) WTW simulation assuming Japan growth slows down and stagnates. The major trading nations such as the USA and China are impacted—and their GDP growth is flat.
(c) WTW simulation assuming Israel growth spikes down by 20 %. The major trading nations such as the USA and China are not impacted—and their GDP growth is steadily rises as if nothing happened.

6.10 Securing the Supply Chain

Given the importance of trade to the economic health of a nation it is perhaps no surprise that securing trade routes and supply chains is at the top of the priority list for major traders like the US. The British Empire used naval power to guarantee delivery of goods to the British Isles and naval power continues to be a critical link in maritime security. Robust supply chains provide natural resources such as oil, agricultural products such as corn, and manufactured products such as iPads and automobiles to global markets. Global power equals global supply chain power, and supply chain power comes down to maritime domain superiority.

Lieutenant Commander Robert Keith was a US Coast Guard inter-agency liaison officer and program manager responsible for identifying and coordinating global security efforts in 2012. His job was to coordinate 29 US government agencies, 4 allied country counterparts, and numerous regulatory bodies, various non-governmental organizations, and the maritime commercial sector to secure the global maritime supply chain.

Figure 6.8 shows a piece of Rob Keith's area of responsibility. It is a portion of the physical WTW servicing import/export of goods through the Los Angeles/Long Beach port. LA/LB is one of the mega-hubs we all depend on for automobiles, computers, food, pharmaceuticals, etc. The combined ports of Los Angeles and Long Beach rank number seven in the world in terms of TEU (Twenty-foot Equivalent Unit, a measure of container capacity), after Shanghai, Singapore, Hong Kong, Shenzhen, Busan, Ningbo-Zhoushan, and slightly ahead of Guangzhou,

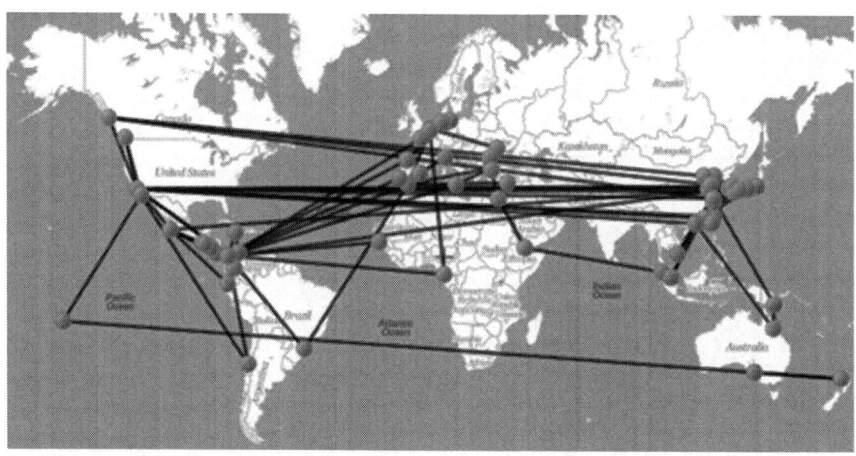

Fig. 6.8 Segment of the world trade web connecting the LA/LB port with the rest of the world shows dependencies with Panama and other ports

Qingdao, Dubai, and Rotterdam. LA/LB port authority employs 30,000 people and moves $100 billion/year of cargo in and out of the US.

Only two other US ports even rank in the top 50 ports: New York/New Jersey, and Savannah. More importantly, these top-ranked import/export hubs form a tightly coupled trade network as shown in Fig. 6.8. What happens if one of them is blocked, disabled, or shut down for a period of time? We know something about this, because of the 2011 Fukushima dai-ichi tsunami and its impact on the electronics and automotive industries.

6.11 I Want My MTV

The nuclear power meltdown and tsunami caused by the Great East Japan earthquake of 2011 took out much of the supply chain underlying Japanese trade. Now estimated at $500 billion in secondary economic damages, the catastrophe affected companies like Toyota and Apple. For example, Toyota reported a 30 % decline in revenues ($2.5 billion loss), because the company could not get parts for its factories. The impact on the global electronics industry may have been even worse, except for the fact that much of the electronics industry has shifted to Singapore, Taiwan and China. [Singapore and China ports are now the largest in the world, so the electronics supply chain may be only as secure as these mega-hubs].

Apple Inc., for example, depends on the global supply chain to design, manufacture, and distribute its *iPod*, *iPhone*, and *iPad* products. The *iPhone*, for example contains processors from Samsung and broadband communications chips from Infineon via Singapore. Taiwan supplies cameras (Primax), circuits (Foxconn), printed circuit boards (Umicron Tech), connectors (Entery Industries),

Bluetooth chips (Cambridge Silicon), and metal cases (Catcher Tech). China (Foxconn) does assembly, touch screen controls (Broadcom), and 802.11 wi-fi chips (Marvel).

The enthusiastic consumer can go online and track the purchase of an *iPhone* as it moves from Shanghai to Alaska, San Francisco, LA/LB, London, etc., and finally to doorstep over a period of 3–5 days. The chain is highly efficient, optimized, and cost-effective. But, one misstep and this network will be seriously damaged because of its high degree of *self-organized criticality*.[7]

6.12 Ports of Importance

Rob Keith's analysis of the LA/LB port piece of the supply chain suggests a considerable build up of *self-organized criticality* in the supply chain network. If ports and trade routes are ranked according to degree (the number of links connecting a port to the trade network), the top three ports in Commander Keith's analysis are Panama, LA/LB, and Shanghai. If the ranking is according to *betweeness* (the number of paths through a node in the network), the most important assets are Panama, LA/LB, and the trade route between LA/LB and Valparaiso, Chile). Finally, if the criterion for determining self-organized criticality is risk, the ranking is Panama, Hong Kong, and Shanghai. The same ranking holds when combining risk, degree, and betweeness into one measure: Panama, Shanghai, and Hong Kong. Apparently, these three ports are the key to WTW security.

The rank-order exceedence of the international supply chain network conforms to a long-tailed distribution. At the long end are *black swan* critical assets—Panama, Shanghai, and Hong Kong. This is rather obvious in the cases of Shanghai and Hong Kong, because they are so large. It is less obvious for Panama, because it is not actually a port, but instead, a channel. However, the 50-mile Panama Canal is an East-West bottleneck. Its betweeness is very high. Another lane is being constructed next to the 100-year old Panama Canal (it opened in 1914), so that larger boats can pass through. The "third set of locks project" is scheduled to be operational in 2014. But the same amount of risk remains because of the proximity of the old and new locks. Without the Panama Canal, half of the world's trade is in jeopardy.

We also obtain a long-tailed exceedence probability distribution for risk versus consequence when simulating single-port and single trade route failures. But the long-tailed distribution qualifies this trade network as a low-risk hazard, because risk declines with increasing severity of consequences. But tell that to Toyota Motor Company.

---

[7] Recall that SOC is the force-multiplied that increases risk and lowers resiliency because of hubs, betweeners, and other optimizations to a complex system.

6.13 Shocking Results

Wealth steadily increases with volume of trade because of comparative advantage. But nonlinear chaotic oscillations are introduced into the trade network when it is shocked by a natural or human-caused event. These shocks impact nations in different ways depending on their connectivity to the world trade web and financial fitness.

Large economies with a large number of trading partners (connectivity) are resilient against financial shocks from abroad. The opposite is true of small economies with small number of trading partners. Small traders will be rocked by large traders, but the opposite is not true—large traders are protected by numbers. A large deviation in the GDP of Egypt, for example, will have very little effect on the US or Europe, but a small deviation in the GDP of the US or Europe will have a major impact on Egypt.

The wealth of nations has become increasingly dependent on the world trade network and a handful of ports and trade routes. Self-organized criticality has evolved to its current level because the global supply chain has been optimized for performance—not security. And since financial resilience depends on trade, this fragility should be of major concern to nations everywhere.

What kind of shocks should we be prepared for in the future?

References

Chinazzi, Matteo, Fagiolo, Giorgio, Reyes, Javier A., and Schiavo, Stefano (2012), "Post-Mortem Examination of the International Financial Network", (January 30, 2012). Available at SSRN: http://ssrn.com/abstract=1995499 or http://dx.doi.org/10.2139/ssrn.1995499

Feigenbaum, Mitchell J. (1978), "Quantitative universality for a class of nonlinear transformations", *Journal of Statistical Physics*, vol. 19, no. 1, pp. 25–52, 1978. DOI:10.1007/BF01020332

Serrano, M. Angeles, Marian Boguna (2003), "Topology of the World Trade Web", Physical Review E, 68, 015101 (2003), DOI:10.1103/PhysRevE.68.015101

Xtremes 7

Abstract

If the 21st century is punctuated and black swan events are becoming more frequent and larger due to a highly connected world, then what is the worst thing that can happen? Might it all end with an Xtreme big bang? Or, will the end come as a slow death due to climate change or mismanagement of global resources? It seems that humans are capable of recovering from almost anything, except for human nature. In this chapter I examine in careful detail the evidence—pro and con—for global climate change. The evidence is undeniable, even as many policy-makers choose to deny it. The temperature of the earth is rising in concert with greenhouse gases. But, the measurements are messy—nonlinear and chaotic—which brings them into question by naysayers. Don't be fooled by complexity—natural disasters will continue to increase in severity and frequency as the weather is influenced by human activity. The most extreme events lie ahead of us.

7.1 The Big One

This is a true story. A team of astronomers at JPL recently detected an object three times the size of a football field headed right towards planet Earth. The huge rock's trajectory was supposed to intersect with the earth's path by 2036, and slam into the ground (or ocean) with a force equal to a 500-megaton atomic bomb. The largest atomic bomb ever exploded on earth, "Tsar Bomba", was a 50-megaton bomb detonated by the Former Soviet Union in 1961. In comparison, this asteroid would be ten times as big. The astronomer's concern showed in the name they eventually assigned the asteroid—*Apophis*—the doomsday asteroid. Actually, Apophis is the name of the Egyptian god of darkness known for doing battle with the sun god Ra. Would Apophis blot out the sun? Could it be the Big One?

Some 49,000 years ago a large nickel-iron meteorite half the size of Apophis hit Earth traveling 40,000 miles per hour—equivalent to the explosive power of a 20-megaton bomb. It left a 575-foot by ¾ mile tourist attraction in the Arizona desert called *Barringer Crater*, named after its 1902 discoverer. Luckily, Barringer Crater is in the middle of an unpopulated desert. If Apophis lands on New York, it could wipe out the entire island of Manhattan along with its 8 million day trippers.

Jon Giorgini, Lance Benner, and Steven Ostro of CalTech's Jet Propulsion Labs, Michael C. Nolan of the Arecibo Observatory, Puerto Rico, and Michael W. Busch of the California Institute of Technology first spotted Apophis in 2004. Their preliminary data suggested a relatively high probability of colliding with earth. Images subsequently captured by a European space telescope revealed it to be much bigger than previously thought, but veering away from its fatal rendezvous with people of earth. The likelihood of a collision with the doomsday asteroid was re-computed, and now it has one chance in 45 of landing on Earth. A 2012 study predicted that Apophis will pass within 22,364 miles of New York in April 2029. Scientists continue to update estimates. As Apophis gets closer to earth, who knows how close it will come to crashing into the globe?

Apophis isn't the last asteroid to threaten earthlings. It isn't even the worst thing we can imagine. But near misses like Apophis raise the existential question, "what could end life on earth?" Asteroid attacks are so extreme it is difficult to contemplate their inevitable impact. A relatively small asteroid might not demolish the planet, but it could destabilize the climate so much that life as we know it can no longer survive here. It has happened before.

7.2 The End of the World: Again

Scientific evidence continues to mount: life on earth has been threatened so many times before that reports of existential near-misses have become rather routine. In terms of the vast geological clock, near misses are a way of life—or death. For example, the *Carboniferous Rainforest Collapse* (CRC) 305 million years ago sent earth into a short ice age that eliminated dense vegetation in Europe and America. CRC created vast coal seams that we now burn in our power plants. Some scientists claim CRC was the result of very large volcanic plumes that blotted out the sun. Others say it was caused by a swarm of asteroids attacking all at once.

An even more serious event called the *Permian-Triassic Extinction* (*Great Dying*) occurred 252 million years ago, and nearly ended us before we had a chance to begin. The Great Dying wiped out 96 % of marine animals and 70 % of terrestrial life. It took 10 million years to recover! Again, the calamity was due to an abrupt climate change—a sudden release of methane, which is a greenhouse gas (*GHG*). These gases are thought to have emanated from fires and volcanoes. The earth must have been a burning inferno 252 million years ago.

The earth went through an extreme warming period about 55 million years ago, perhaps also due to GHG release. More recently, scientists using archeological traces found in rocks, soils, and trees, have evidence of just the opposite—abrupt

7.2 The End of the World: Again

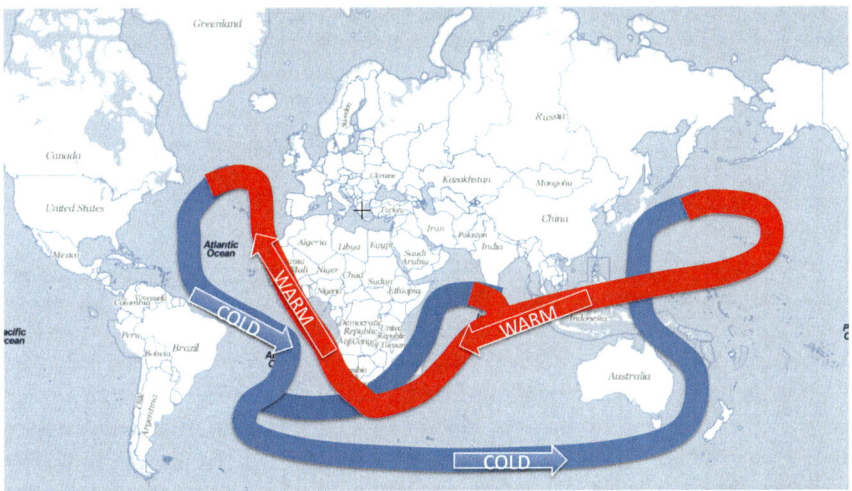

Fig. 7.1 The Atlantic conveyor is part of the global circulation system of water currents called *Thermohaline Circulation*—the movement of oceans from the Southern Hemisphere to the North. Warmer water flows from the Indian Ocean, around South Africa, up the Gulf Stream bordering Eastern US, to Greenland. It distributes its heat to North America and Europe, and then cools in the North Atlantic. The cooler water sinks below the surface, and returns back to the Antarctic and Indian Ocean, where the cycle repeats. NASA/JPL

cooling. The *Younger Dryas Stadial* event 12,000 years ago is known as the *Big Freeze*, because immediately after the event, earth's temperature suddenly dropped. This cold spell lasted for 1,300 years, and established the Siberian land bridge that opened the door between Asia and North America allowing humans to drift into the Western Hemisphere.

The Big Freeze might have been caused by a shift in the jet stream or an extreme solar flare, but a more likely explanation is that spillage from Lake Agassiz stalled the *meridional overturning circulation* (MOC) system of ocean currents that warms the Northern Hemisphere. MOC is a type of *thermohaline* circulation system that moves water and heat from the southern to northern hemisphere. But MOC and thermohaline circulation are mouthfuls, so many people call the North American thermohaline circulation system the *Atlantic conveyor*, because it works like a heat conveyor belt moving warm water northward, and cold water southward, see Fig. 7.1. The Atlantic conveyor controls the temperature of the entire planet much like a thermostat controls a room's temperature.

Lake Agassiz was a giant body of fresh water covering most of Canada. For some reason it spilled over into the Saint Lawrence River and Seaway and then flowed out to the North Atlantic. Here it met with the Atlantic conveyor and either temporarily stopped it or slowed it down. Thus, the earth's heating and cooling system was disrupted. [Thermohaline circulation depends on salinity, sunlight, and winds, so movement of heat is not a simple relationship—it is a complex system].

Earth's temperature has never traveled in a straight line. It seems to oscillate from one extreme to another. *Dansgaard-Oeschger (D-O) Events* are oscillating warming episodes lasting a few decades followed by longer cooling periods. The last one was 11,500 years ago. Greenland warmed by 4 °C within a relatively short 40 years, and then cooled down only to warm up again.[1] D-O Events may be related to *Heinrich Events*—huge volumes of fresh water flowing into the ocean, which decreases salinity, which in turn cools the ocean. Since most of the heat stored on earth is in the oceans, small changes in the *SST—sea surface temperature*—makes a big impression on global temperature.

Large sheets of ice, called glaciers, can also change the earth's temperature by melting and freezing. *Bond Events* are warming-cooling oscillations that occur about every 1,470, plus or minus 500, years. They mostly happen in the North Atlantic. Nine cycles have been observed so far. Fred Singer and Dennis Avery report, "The Earth has been in the Modern Warming portion of the current cycle since about 1850, following a *Little Ice Age* from about 1300 to 1850. It appears likely that warming will continue for some time into the future, perhaps 200 years or more, regardless of human activity."[2] The contemporary global warming episode is not the first warming episode to affect climate.

And then there are the Milankovich cycles—climate changes occurring every 21,000 and 41,000 years because the earth wobbles on its axes. Serbian geophysicist and astronomer Milutin Milanković worked out the cycles while a prisoner of WWI. Currently, we are in the early stages of a cooling cycle.

One theory of global warming connects glacier contraction and expansion with the Atlantic conveyor, which rules over global temperatures. Another theory is that climate change is a byproduct of complex system interactions among forcing functions such as the sun, ocean currents, earth's magnetic field, tectonic motion, etc. There remains plenty of work for climatologists to do to figure out how this complex system works. In the meantime, climate change is a controversial topic.

We know one thing: extreme shifts in the earth's ecosystem can lead to warming and cooling cycles, and in the most extreme cases, extinctions. A collision with an errant asteroid isn't needed to end it all here on earth. Escalating superstorms and melting glaciers can do what asteroids can't! Could global climate change be the Big One that ends it all—again?

7.3 The Carrington Event

Early in the morning of September 1, 1859, amateur astronomer Richard Carrington cranked open the dome of his private observatory to scan the bright London sky. The sun's spots attracted his attention, so he aimed his 19th century

---

[1] Climatologists use Celsius instead of Fahrenheit to measure temperature. 4 °C is 39 °F. Water freezes at 0 °C, which is the same as 32 °F.

[2] Singer et al. (2005).

7.3 The Carrington Event

telescope at the sun and began sketching what he saw—"two patches of intensely bright and white light" erupting from the sunspots.[3] The flares vanished almost as suddenly as they appeared, but within hours, their impact was felt across the entire globe. Later that day, telecommunications (telegraph) began to fail around the world. Some machines showered operators with sparks and set papers on fire. Bright and colorful aurora borealis "light shows" appeared all over the planet. Some birds and people thought it was the beginning of a new day and set about chirping and going to work in the middle of the night. Other animals and people thought it was the Big One and began preparing to meet their creator.

The Carrington event was disconcerting if not damaging,

"The sky was so crimson that many who saw it believed that neighboring locales were on fire. Americans in the South were particularly startled by the northern lights, which migrated so close to the equator that they were seen in Cuba and Jamaica. Elsewhere, however, there appeared to be genuine confusion. In Abbeville, South Carolina, masons awoke and began to lay bricks at their job site until they realized the hour and returned to bed. In Bealeton, Virginia, larks were stirred from their sleep at 1 a.m. and began to warble. (Unfortunately for them, a conductor on the Orange and Alexandria Railroad was also awake and shot three of them dead.) In cities across America, people stood in the streets and gazed up at the heavenly pyrotechnics. In Boston, some even caught up on their reading, taking advantage of the celestial fire to peruse the local newspapers."[4]

Carrington observed the largest known solar flare—a massive solar explosion with the energy of 10 billion atomic bombs. Traveling at the speed of light across the 93 million miles between the sun and earth, the flare eventually reached earth where its electrified gas and subatomic particles produced a *geomagnetic storm*—dubbed the *Carrington Event*. Geomagnetic storms contain highly charged EMP (Electromotive potential) particles that play havoc with electronic systems. In 1883 this meant telegraphy. But in 2012 messing with electronics means messing with just about everything we depend on for modern life. A Carrington Event would wipe out the Internet if it happened today.

A 2008 report from the National Academy of Sciences claims it could cause "extensive social and economic disruptions" due to its impact on power grids, satellite communications and GPS systems.[5] Power grid engineers are concerned that solar radiation might melt the copper wires in power lines, and scramble circuits in control computers. The impact could be enormous because modern life depends on electrical power to run the Internet, transportation systems, factories, offices, and food and water supply systems. For example, a relatively small geomagnetic storm in March 1989 left six million people in Quebec, Canada without power for 9 hours.

---

[3] http://www.thetruthdenied.com/news/2012/12/10/geomagnetic-storms-and-their-impacts-on-the-u-s-power-grid-pole-shift/
[4] http://www.history.com/news/a-perfect-solar-superstorm-the-1859-carrington-event
[5] National Academies Press (2008).

While scientists have been measuring earth's evolving and complex nature for perhaps hundreds of years, I consider the Carrington Event the beginning of *climatology*—the study of earth's climate. It embraces elements from atmospheric sciences, physical geography, oceanography, and biogeochemistry. Moreover, Carrington's hand-drawn sunspot sketches marks the beginning of a new idea—that earth is part of an ecosystem that extends beyond land, sea, and air. Climatology is about systems and how they interact in complex ways. And as earth's ecosystem evolves, it changes the very life it spawned. Humans are a product of this ecosystem. As it changes, so must we.

7.4 Black Bodies

A few decades before Carrington discovered sun spots from observations made using his personal telescope, a Scottish engineer named William Thomson was knighted for his role in laying the first transatlantic cable connecting Europe with the new world. Sir William Thomson, Baron Kelvin of Largs, Lord Kelvin of Scotland (1824–1907) is better known for his breakthrough paper, *On An Absolute Thermometric Scale*—an even greater contribution to science. Lord Kelvin calculated the temperature of the coldest thing in the Universe, and established the yardstick used today to measure global temperature. *Absolute zero* is the temperature of matter when not a single molecule moves. Zero is -273 °C to you and me. It is 273 °C below the temperature of frozen ice. One degree K is equal to 1 °C, because measurements begin at zero in the Kelvin scale, denoted by °K.

All life on earth depends on radiant energy from the sun as shown in Fig. 7.2. That energy is quickly turned into heat to keep the oceans circulating, the air cycling, and plants and animals growing. Accordingly, the earth must maintain a balanced *energy budget*—an accounting system whereby heat arriving on earth must also leave, because if any heat is trapped here on earth, it contributes to a continuous rise in temperature. And since terrestrial life has adapted to temperatures in a relatively mild range centered on 288 °K, any departure from this sweet spot is considered a threat to all life. Simply put, all living things on planet earth depend on a fragile balance between heat coming in and heat leaving the planet. Earth's heat budget must be balanced or else the earth will either freeze or boil out of existence.

The famous *Stefan-Boltzmann Law* relates heat to Kelvin temperature. It says the amount of heat radiated by a completely black ball of matter is proportional to its temperature raised to the fourth power.[6] That is, a black body radiates heat proportional to T^4. Therefore, in Fig. 7.4 radiant sun energy arrives on earth containing 341 watts per square meter (W/m^2), of which 102 W/m^2 immediately bounces back into space. The remaining heat continues through the atmosphere and hits the ground or ocean where 23 W/m^2 are reflected again. This leaves 161 W/m^2 to power life on earth.

---

[6] Heat $= \sigma T^4$, where σ is Stefan's constant, 5.67×10^{-8}, and T is temperature measured in K.

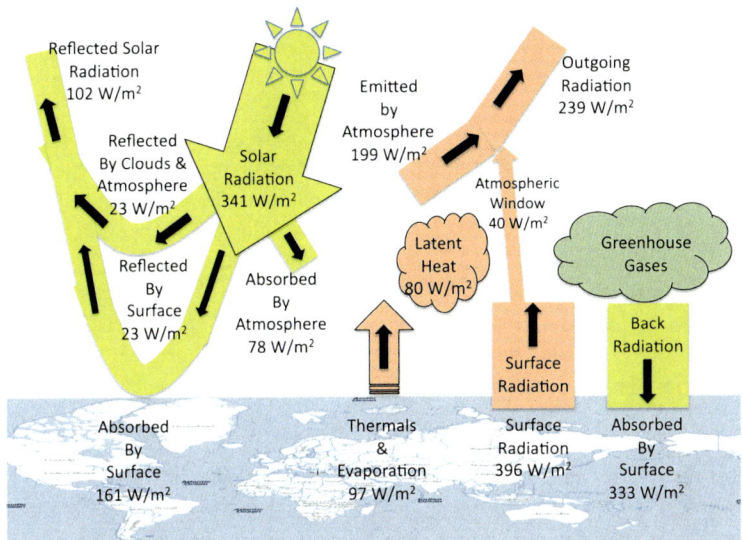

Fig. 7.2 A simple model of the earth's heating and cooling system sometimes referred to as the earth's energy budget. NASA/JPL (Trenberth et al. 2009)

The earth is considered at *gray body* instead of a perfect black body, because much of sun's energy bounces off the atmosphere, clouds, land, and sea. The process of reflecting heat is called *albedo*, and the fraction of energy reflected by a body is called *emissivity*. The average emissivity of earth is about 39 %. This means only 61 % of sun's energy reaches earth. Even after the earth's continents and seas are heated by the remaining 61 %, earth temporarily stores accumulated heat and then radiates it back into space, according to the Stefan-Boltzmann Law. When earth's energy budget is in balance, heat in equals heat out, and life goes on.

If the sun were to stop shining, earth would eventually return to absolute zero. If the sun's radiation increased, as it does during solar storms, earth's temperature rises until another equilibrium temperature is reached and the Stefan-Boltzmann Law kicks in and radiates heat back into space to maintain equilibrium. If the sun's rays remain unchanged, earth's heat budget remains in equilibrium between heat in and heat out. The earth's temperature remains steady under equilibrium conditions. Life on earth has adapted to this temperature equilibrium or *fixed point*.

If earth were simply a black body absorbing all 341 W/m² of incoming radiation without albedo, its average temperature would be 279 °K (6 °C; approx 43 °F). But allowing for albedo, earth's temperature would reach equilibrium at 255 °K (−18 °C; approx 0 °F). A cold planet, indeed!

But earth's pre-industrial average temperature is generally agreed to be considerably higher at 288 °K (15 °C; approx 59 °F). Why the difference? The answer is that GHG (greenhouse gases) like water vapor, methane, and CO_2 provide additional warming. In other words, life on earth *depends* on GHG. Life as we know it is possible because GHG makes earth warmer and establishes an equilibrium at a mild 59 °F.

This process is shown in the right-hand side of the energy budget diagram of Fig. 7.2. As you can see, back radiation from GHG supplies an additional 333 W/m^2 of heat back to earth. But most of it is absorbed back into the atmosphere through a variety of complex heat transfers between land and sea, sea and air, and more black body radiation. When earth is in balance, the net-net energy budget is zero, which stabilizes global temperature. However, when the budget is not in balance, earth's temperature either rises or falls, depending on the type of heat transfer taking place.

7.5 Models of Global Warming

The key to understanding global warming is in Fig. 7.2, but the underlying mechanism driving the processes in Fig. 7.2 is not well understood. This is why climatology is still a young science. Climate modeling is the first step in understanding the mechanisms of any complex climate system. So a number of climatology models have been developed and back-tested to validate them against historical data. Admittedly, the evidence is more empirical than theoretical, so a certain amount of skepticism is warranted. But it is a start.

The first question most people ask is, "how do we know the earth is warming up?" That is, how do we know there is an imbalance in earth's energy budget? To answer this question we can study temperature anomalies recorded over the past 100 years or more. *Anomalies* are departures from recent historical measurements of temperatures taken throughout the year and across the globe. Then we can compare recent measurements against previous years' measurements to determine the size of anomalies and determine any trends.

Figure 7.3 plots temperature anomalies obtained from the HADSST2 data set. The HADSST2 Sea Surface Temperature anomaly dataset is maintained by one of the most trusted and respected science centers in climatology—the UK Meteorological Office Hadley Center.[7] Its collection is sophisticated, and sometimes controversial, however [The controversy is addressed later].

Sea surface temperature (SST) readings are collected from nearly 2600 latitude and longitude squares beginning at the International Dateline and the North Pole and moving eastward. The measurements are anomalies, not absolute values. For example, an anomaly of 0.5° means the temperature at that square is one-half of a degree above an historical baseline value. Typically, the baseline is an average calculated in 1960, 1980 or perhaps 1880 (pre-industrialization). Figure 7.3 shows measurements made since 1980.

What about Fig. 7.3? First, note that the data are noisy—readings are spread all over the graph and appear to be random. In fact, there is a non-zero probability that a straight-line regression as shown in Fig. 7.3 is fooled by randomness. But let us

---

[7] http://hadobs.metoffice.com/hadsst2/index.html

7.5 Models of Global Warming

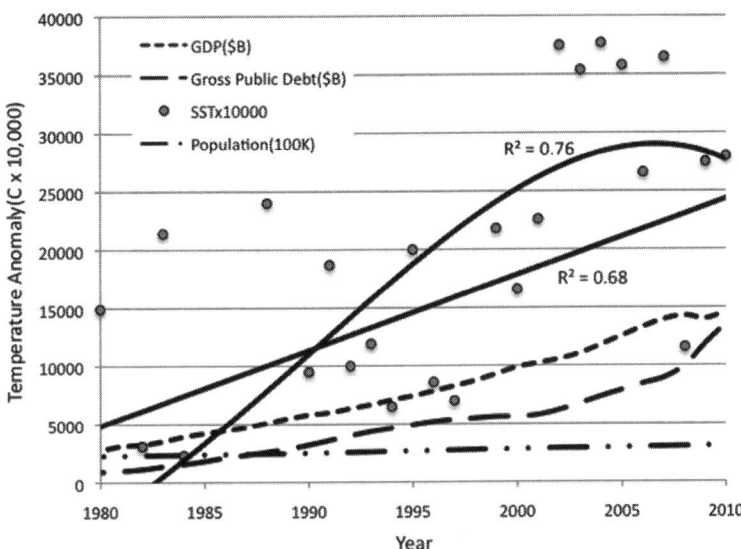

Fig. 7.3 Sea surface temperature anomalies are compared with USA GDP, USA Public Debt, and USA population. The raw data is scaled by a factor of 10,000 and then two least squares curve fits are drawn: a straight line with correlation of 0.68 and a polynomial of order 6 with correlation of 0.76

assume an upward trend exists. The correlation coefficient of the straight line is modest (0.68), suggesting a poor fit. What happens if we fit a polynomial to the random-looking anomalies? The best fit made possible by Microsoft Excel is a polynomial of order 6, which better matches the data (0.76). In addition, the polynomial shows a reversal of the decade long upward trend after 1998—the hottest year on record.

An objective scientist cannot conclude that global warming is on the increase from Fig. 7.3 alone. In fact, claims that global warming is caused by an increase in population is also difficult to support from Fig. 7.3, alone. Is warming caused by an increase in the USA population? Is it caused by the GDP of the US? By the US public debt? All of these are increasing, but GDP and Public Debt show a stronger correlation than population. The only conclusion we can draw from Fig. 7.3 is that correlation is not the same as causation.

7.6 Anomalies in the Anomalies

Empirical data analysis isn't conclusive enough to prove either global warming or cooling. Perhaps the 21st century is nearing the end of a Bond Event, as suggested by Singer and Avery. Or perhaps the empirical data ignores other factors such as a bias toward urban temperature measurements in the data. Indeed, HADSST2 temperature measurements have been criticized from a number of angles. Sensors are not evenly distributed over the surface of the earth—69 % of them are in the 30–60° latitudes, so there is a northern hemisphere bias. Almost ½ are in the US, so there is a USA bias. Second, the raw data is periodically adjusted to account for a number of factors such as upgrades to thermometers and changes in procedures. Since 1960, these adjustments have raised anomaly measurements by nearly 0.6 degrees.[8]

And then there is the complex dynamics of interactions among the oceans, air, and bodies of land. For example, the ocean's ability to absorb heat (or not) is poorly understood. There may be lags between heating and measuring. And as noted above, small changes in global thermohaline circulation can dramatically alter temperatures around the globe. The thermohaline conveyor is easily disrupted. Perhaps temperature anomalies are the result of nonlinear chaotic complex processes. Perhaps the rapid rise since 1980 is a nonlinearity that will soon dissipate. Many alternative explanations for global warming are possible.

7.7 Global Warming Isn't Linear

If global temperature change is truly a symptom of a more complex ecosystem, then we should be able to measure and observe chaotic twists and turns in the data of Fig. 7.3. For example, the data may look like random noise because it contains nonlinearities and chaotic oscillations. Or, the data may simply be so noisy that the signal is lost in the noise. To study the "nonlinearity versus noise hypothesis", I constructed a series of state-space diagrams at various resolutions as shown in Fig. 7.4. If chaotic oscillations exist in temperature anomaly data, the oscillations should show up as twists and turns in state-space. If a trend exists, it should show up as a trajectory moving up or down. If the data are simply noisy, the state space diagram should look like random strokes in a Jackson Pollock painting.

Recall that a state-space diagram is simply a plot of time-series data against itself, but shifted by one time period. This is what I did with the HADSST2 data of Fig. 7.3. Figure 7.4 plots temperature anomalies in year t against the same anomalies in year (t + 1) for three different resolutions: 10-year average, 5-year average, and one measurement taken in January of each year. Specifically, Fig. 7.4a plots decade averages against one another—decade 1900 versus decade 1910, 1910 versus 1920, ... 2000 versus 2010. Figure 7.4b plots semi-decade averages over 5-year periods: 1900, 1905, 1910, ... 2005, and 2010. And Fig. 7.4c plots single points in 1900 versus 1901; 1902 versus 1903, etc.

How might the state-space diagrams of Fig. 7.4 be interpreted? First, even at the coarse 10-year resolution, HADSST2 anomaly data contains a non-linearity. The loop shown in the center of the figure—around 1940–1960—rules out any

---

[8] www.appinsys.com/globalwarming

7.7 Global Warming Isn't Linear

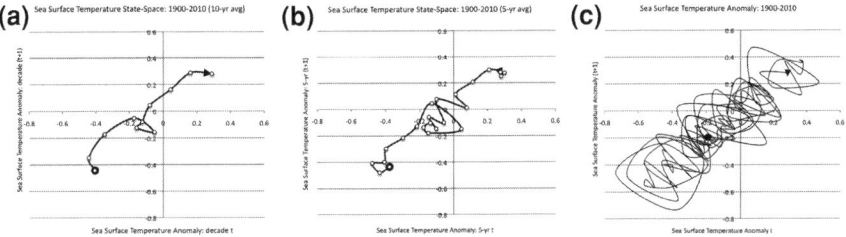

Fig. 7.4 State-space diagrams of sea surface temperature anomalies show signs of nonlinear chaotic behavior. Cyclic behavior indicates nonlinearities, and fractal self-similarity suggests the likelihood of underlying chaotic processes. **a** 10-year resolution. **b** 5-year resolution. **c** Annual resolution

thought of fitting a straight line to temperature anomaly data. While it is not clear what causes the nonlinearity, an objective scientist cannot accept the hypothesis that HADSST2 temperatures obey a linear relationship across the years. A more convincing conclusion is just the opposite—HADSST2 measurements were obtained from a nonlinear process of some sort.

Second, the repeated loops that emerge as the state-space diagram is scaled from decade to semi-decade, and finally annual measurements are somewhat random, but self-similar patterns—another indicator of chaotic processes underlying the phenomena. The high resolution (1-yr) and low-resolution (10-yr) state-space diagrams contain similar patterns, but at different scales or resolutions. [Think of Fig. 7.4a as a photograph of Fig. 7.4b taken from an altitude of 10,000 feet. Detail is lost at this distance, so detailed twists and turns are obscured. But as we increase scale and move closer, greater detail materializes, until the state-space diagram is full of twists and turns, per Fig. 7.4c. Remember the ocean waves at Asilomar Beach described in the first chapter on Waves?]

7.8 Global Warming is Chaotic

Temperature anomaly noise can be filtered out or smoothed over to get at trends using more sophisticated techniques than presented here. What is left after smoothing should give a truer picture of the direction of global temperatures. In fact, two Chinese scientists, Zhen-Shan and Xian, removed noise from temperature anomaly data and observed what was left over.[9] They found two trends as shown in Fig. 7.5, and noted oscillations similar to the oscillations shown in Fig. 7.4 as well as long-term cycles in temperature data collected over China. Their result is based strictly on statistical analysis, and ignores the physical science that may be forcing HADSST2 temperature to oscillate.

---

[9] Zhen-Shan and Xian (2007).

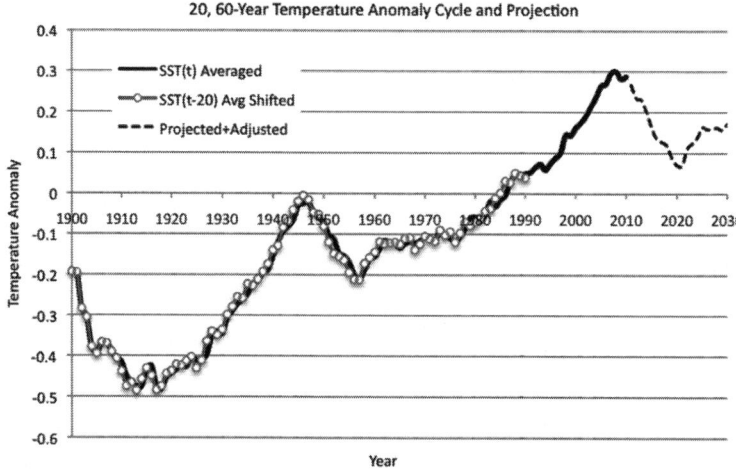

Fig. 7.5 Temperature anomaly measurements used by scientists to forecast global warming to 2030 contain 20 and 60-year cycles. Future temperatures repeat the pattern of 60-years earlier, according to researchers Zhen-Shan and Xian

Using data going all the way back to 1881 through 2002, and much more sophisticated statistical techniques than considered here, the two scientists from the Nanjing Normal University in China concluded, "temperatures have been decreasing since the year 2000." More importantly perhaps, is their use of a technique for filtering out noise to get a better look at trends. I won't go into the details, but their method identified two dominant cycles: one of 20 years in length, and another of 60 years.

Using the same HADSST2 data going back to 1900, Fig. 7.5 shifts decadal average values (smoothing out noise) by 20 years to obtain an incredibly accurate match with temperature anomalies from 1900–1990. That is, decadal averages from 1920–2010 are superimposed on decadal averages from 1900–1990. The match is so exact that it is impossible to distinguish between the two time-periods plotted in Fig. 7.5.

Figure 7.5 uses decadal averages from 60 years in the past—1950–2010—to forecast the next 20 years of temperatures. The pattern repeats, except an increase of 0.28 °C must be added to projections beyond 2010. Note how temperature decreases according to the Zhen-Shan and Xian prediction, but remain higher than the previous 20-year cycle. Accordingly, temperatures will drop over the next 20 years and then begin to climb again as the 20-year cycle starts over. Unless of course, underlying mechanisms of climate dynamics are altered.

Even after chaotic oscillations have been removed, a temperature anomaly of roughly one-quarter of a degree exists since 1980. That is, the earth is getting warmer, even though its temperature path contains ups and downs. But, there is no linear trend. Rather, two cycles exist in the data—a 20- and 60-year cycle—that both trend slightly upward.

Is this upward trend a fact of nature or is it caused by human development? Does the gradual rise in temperature correlate with human activity, or is it due to a much larger influence—earth and the solar system, itself?

7.9 The Lightning Rod

James Edward Hansen has been called a hero for advocating *anthropogenic* (human-caused) global warming, and also criticized for overestimating its dangers.[10] He is perhaps the staunchest spokesperson warning the rest of us about global warming, having received numerous awards for his 40 years of climatology research and "courageous and steadfast advocacy in support of scientists' responsibilities to communicate their scientific opinions and findings openly and honestly on matters of public importance."[10]

But physicist Freeman Dyson criticized Hansen in 2009 for exaggerating global warming. According to an interview in the *New York Times*, Dyson complained that, "Hansen has turned his science into ideology." In 2011 Hansen was arrested while protesting in front of the Whitehouse. He has become a lightning rod in the debate over whether climate change is caused by industrialization or is simply another Bond Event.

Hansen began his journey to celebrity as a 1960s and 1970s student of the Venusian climate. Venus' climate may have been similar to earth's several million years ago. But a *runaway greenhouse effect* evaporated its water and turned the second planet into an uninhabitable hothouse. Hansen worries that earth's fate may be the same. Furthermore, he believes fervently that the future of earth's climate is in humanity's hands. We can ruin the planet or save it, depending on our policy choices.

Hansen says people can prevent another runaway greenhouse disaster by shutting down coal-burning power plants, adopting energy efficiency policies, switching to renewable energy, converting our old dumb power grid to a smart grid, and producing electricity using fourth-generation nuclear reactors that burn nuclear waste rather than produce it. The science is easy, but the politics are difficult.

When asked about how long we have before earth's ecosystem is ruined forever, Hansen said, "It is difficult to predict time of collapse in such a nonlinear problem."[11] But he maintains that collapse is inevitable, unless we change our ways. And, he has solid science to back up his claims, which makes him a credible and powerful spokesperson for global-warming-as-catastrophe.

Hansen's scientific publications are characterized by lucid simplifications of complex ideas. They show a deep understanding of nonlinear complexity, but they are rather simplistic and obvious. Take the Hansen-Sato Green's function as an

---

[10] http://en.wikipedia.org/wiki/James_Hansen
[11] http://en.wikipedia.org/wiki/James_Hansen

example—it captures the essence of the problem, and it is so easy to understand. This simple model says that global warming is simply the sum of a handful of forcing functions—causes of temperature anomalies.

7.10 Green's Forcing Function

Hansen and Sato et al. proposed one of the most fundamental models of climate dynamics in an online paper published in 2007.[12,13] Their model makes it easy to understand climate dynamics because it breaks the problem into parts called *forcings*—causes-and-effects that either lower or elevate earth's temperature. *Forcing functions* are simply sub-models that represent the impact of a physical process like CO_2 loading on another process like rising temperature. Once each forcing function sub-model is understood, the overall "big climate picture" can be constructed by summing all *forcings* together. The summation is a type of *Green's function*, hence the name of the model.[14]

Climate forcings can be caused by CO_2, which increase temperatures, and volcanic aerosols that decrease temperatures. The major GHG forcings are water vapor from the oceans, methane from agriculture, and CO_2, from a variety of sources, including industrialization. Temperature anomalies arise because these forcings get ahead (or behind) one another. Each forcing may induce nonlinear oscillations in processes such as heating, thermohaline circulation, solar flares, etc. They may also be related to increased storm intensity—*superstorms*—and other side effects like the extinction of species, and increases in floods and droughts.

The net change in temperature due to all forcings is calculated by substituting forcing function sub-models into Green's function:

$\Delta T = \Sigma R(\Delta F)$
ΔT: temperature anomaly; rise or fall of temperature
ΔF: forcing anomaly or event, e.g., a solar flare or drop in CO_2 concentration
R: response function; how temperature responds to a forcing event
Σ: sum over all positive and negative forcings—solar, aerosols, CO_2, etc.

For example, the IPCC (Intergovernmental Panel on Climate Change) uses a logarithmic sub-model of CO_2 forcing. Assuming a pre-industrial CO_2 concentration of 280 ppm, the increase in temperature anomaly due to CO_2 forcing can be calculated as follows:

$\Delta T = R(\Delta F)$
$R(\Delta F) = 5.35\log(C/280)$
where C is CO_2 concentration.

---

[12] Hansen et al. (2012a).

[13] Hansen et al. (2012b).

[14] Green's functions are used in physics and statistics. They are various types of correlation functions such as the correlation between global temperature and GHG.

7.10 Green's Forcing Function

IPCC models often assume a doubling of CO_2 over the next few decades. If this happens, C in the equation above is set to 560, which means $\log(560/280) = \log(2)$, or 0.7. Hence, the carbon-doubling assumption so frequently cited in the literature means global temperatures will rise by $5.35(0.7) = 3.75$ °C. This does not seem like much, but it could cause sea levels to rise by as much as 20 feet, which would flood many densely populated areas around the world.

As noted in the previous section and Fig. 7.4, temperatures have raised 0.8 °C since pre-industrial days (280 ppm), so according to the IPCC sub-model, the temperature will rise almost another 3 °C over the next century. This is inevitable of course, unless temperature anomalies are nonlinear, cyclical, and chaotic, instead of linear and smooth. The projection also assumes CO_2 forcing increases temperatures according to the IPCC sub-model.

Other forcings are also going on at the same time. Water vapor, nitrous oxide, halocarbons, ozone, volcanic aerosols, and methane all force climate change through imbalances in earth's energy budget. For example, methane and other aerosols can lead to cooling (*negative forcing*) of up to 1.6 W/m². Volcanic plumes cause brief, but deep cooling, and the ocean absorbs both heat and CO_2.[15] Significant but brief negative forcings, have been recorded immediately following major volcanic eruptions including, most notably, the Krakatau eruption in 1883, El Chichon in 1982, and Pinatubo in 1991. The combination of increases and decreases due to various forcings makes temperature prediction difficult and controversial.[16]

In fact, there are so many different forcing sub-models and associated data that even the Hansen-Sato simplification remains inconclusive. Over the years dozens of CO_2 doubling models and assumptions and simplifications have been used to predict temperature anomalies ranging from 2–10 °C. Rebecca Lindsey posts that when these predictions are aggregated into a long-tailed probability distribution, the rise in temperature is more likely to approximate 2–3 °C, with rare black swan increases as high as 10 °C.[17] But even Lindsey's more modest expected value is controversial.

The numerical data in Fig. 7.3 may be contested, but it is difficult to dispute that temperatures are rising, see Fig. 7.5. Is warming anthropogenic or natural? Is the current trend short-lived or permanent?

Whether people, periodic Bond Events, or 20-year natural cycles are the cause of global warming, the inevitable consequences remain a big concern. If global warming really does upset the heat budget, thermohaline cycles, or ice caps, how bad can it get?

---

[15] Nature has absorbed 55 % of all GHG: plants absorbed 2/3, while oceans absorbed the remaining 1/3.

[16] Gregory and Foster (2008).

[17] Rebecca Lindsey (August 3, 2010).

7.11 Consequences

Regardless of whether Hansen's warnings come true or not, a prudent species would be wise to prepare for the worst. Otherwise, the Big One may be a black swan we cannot recover from. What if nothing is done to repeal projected rises in temperatures around the globe? The IPCC report warns of disastrous consequences:

> Warming lakes will reduce water quality
> Heat mortality will increase
> Diseases will spread more widely
> Droughts will induce water shortages
> Crop failures will lead to starvation
> Warming oceans will kill species and reduce food supplies
> Rising sea levels will cause flooding
> Higher coastal waves will cause tsunami damage
> Intensified storms (super-storms) cause widespread damage

The IPCC forecasts a future based on a number of assumptions that may not come true. For example, their climate model assumes a rise in CO_2 of 1 % per year for the next 70 years to achieve CO_2 doubling. It also ignores negative forcings, and assumes global population rises to 11.4 billion people by 2100. Industrialization is blamed for the rise in CO_2, but the IPCC's assumption of 2.3 %/year global economic growth—and its corresponding industrialization—may be too aggressive. [GDP growth rates of the US, 1947–2012, have averaged 3.2 %; global growth rates, 1999–2011, have averaged 3.7 %].

The IPCC largely assumes linearity and ignores possible tipping points or black swans like a series of Krakatau-sized volcanoes, unexpected rises in aerosols, etc. Finally, nobody knows the effect of "deep-ocean mixing", whereby deeper layers of the ocean absorb heat and CO_2 to balance earth's energy budget. What if absorption of heat into the ocean is delayed by 60 years? Might the 60-year cooling cycle be due to this deep-ocean mixing?

If a more robust model incorporating all of the above were used to predict the climatic future, the assumptions missing from the IPCC model may actually reverse the predicted rise in temperatures. It might even predict a *colder* climate, not a warmer one! Is the IPCC being alarmist or prescient? What evidence is there to support an impending climate doomsday?

7.12 Solar Forcing

Historical data supports the IPCC claim that storms are more severe now than in the past. In fact, the word *superstorm* is now part of the English language. Does storm data support this claim? Figure 7.6 plots exceedence probability of hurricane wind speeds during the period 1980–2005 versus an earlier 25-year period 1851–1876. As usual, the data are plotted on a log-log scale to obtain the slopes of

7.12 Solar Forcing

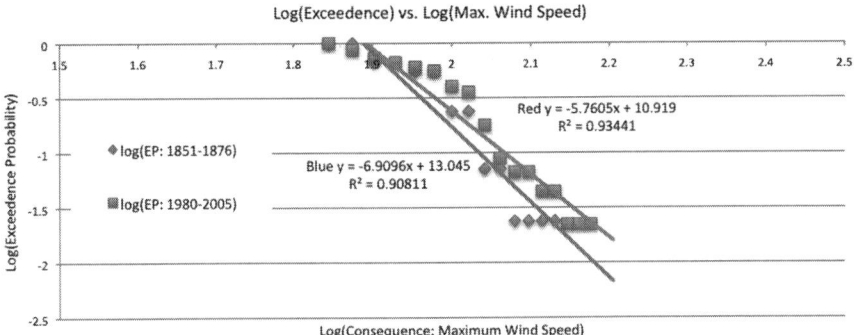

Fig. 7.6 Log-log plot of exceedences for hurricane severity shows an increase in wind speed during recent times versus a century earlier. Correlations are over 90 %. http://www.aoml.noaa.gov/hrd/hurdat/ushurrlist18512005-gt.txt

their long-tailed distributions. The fit is very good for both 25-year periods. [1851 is considered pre-industrial, while 1980 is considered post-industrial]. The 1980–2005 exceedence probability is longer-tailed, indicating more severe hurricanes.

The maximum wind speed of hurricanes recorded in the period 1851–1876 was 130 knots, while the maximum recorded in the period 1980–2005 was 145 knots. The probability of hurricane wind speeds exceeding 100 knots in modern times was 40 %, while the same wind speeds during 1851–1876 occurred with probability 24 %. Thus, the data supports the IPCC hypothesis that severity of hurricanes has increased. There is no reason to believe this trend will reverse.

Why? Climatologists once thought hurricane activity increased with an increase in sunspots. After all, solar flares upset earth's energy budget. But the number of sunspots have historically increased and then decreased on an 11-year cycle. Therefore, hurricane intensity should follow the same 11-year cycle. But synchronization of the number of sunspots with hurricane intensity is not supported by the data. In fact, Robert Hodges and James Elsner, geography professors at Florida State University, claim that solar flares work in the opposite direction relative to the intensity of hurricanes.[18] What matters, according to the two professors, is the difference between sea surface temperature (SST) and high-altitude atmospheric temperature. When the difference is great, and SST is relatively warm, hurricanes are likely to be more severe.

Sea surface and high-altitude temperature differences affect the air-ocean ecosystem. As the difference grows, atmospheric instability grows, which in turn creates more severe hurricanes. If Hodges and Elsner are right, sea surface warming acts like a positive forcing function to energize hurricanes, while sunspot activity acts like a negative forcing function, putting the brakes on turbulence

---

[18] Waymer, Jim (2010).

when sunspot activity warms higher altitudes. It is a contrarian point of view, but many nonlinear forcings contradict common sense.

The Hodges and Elsner model has not been confirmed by other scientists, but their temperature differential theory shows once again the complex relationships between forcing functions and climate change. Some are positive and others are negative forcing functions. In the case of hurricanes, sunspots should pump more energy into the atmosphere and cause more hurricanes. But the opposite appears to be happening. Solar flares reduce hurricane intensity by heating the upper air and reducing the difference between SST and the temperature of the upper atmosphere! This is a counter-intuitive result.

7.13 Other Forcings

Floods, fires, droughts, tornadoes, hurricanes, extreme cold, extreme heat, and just plain severe weather events are on the rise. Table 7.1 lists only those since 1980 leaving behind a bill for damages in excess of $1 billion. Consequences, in terms of cost, obey the familiar extreme event distribution—a long-tailed exceedence probability that fits a power law very nicely, indeed. The probability of future disastrous weather events can be estimated from the a posteriori distribution as shown in Fig. 7.7.

For example, the probability of a major event costing more than $1 billion occurring within the next 300 days is almost 100 % (certain). The probability that it will exceed $30 billion (Superstorm Sandy) is approximately 5.6 %, or about one in 17, see Fig. 7.7a. The probability that fatalities will exceed 100 deaths is 16 % or about four in 21, see Fig. 7.7b. Finally, the probability the US will experience a disaster in excess of $1 billion every month is 22 %, or 2 in 7. In other words, major disasters are the rule, not the exception.

7.14 Pineapple Express

If $1 billion dollar disasters are becoming commonplace then what does a really Big One look like? What does it take for an extreme weather event to be extreme? Superstorm Sandy wasn't anything compared to the Pineapple Express of 1861–62. The truly black swans of weather lie both behind and ahead of us. They have happened before and they will certainly happen again. Pineapple expresses give new meaning to the term, "100-year flood".

In another 20 years, *billion-dollar* superstorms like Sandy (2012) may seem trivial. *Trillion-dollar* superstorms may be the new normal. In fact, superstorms approaching $1 trillion in damages have already happened. Even worse: they are destined to happen again. Figure 7.8 shows a 70 % likelihood of mega-superstorms hitting the West Coast of the US—an estimated probability that will rise

7.14 Pineapple Express

Table 7.1 US disasters in excess of $1 Billion: 1980–2010

Event—Year
U.S. Drought/Heatwave—2012
Western Wildfires—2012
Sandy—2012
Hurricane Isaac—2012
Plains/East/Northeast Severe Weather—2012
Rockies/Southwest Severe Weather—2012
Southern Plains/Midwest/Northeast Severe Weather—2012
Midwest/Ohio Valley Severe Weather—2012
Midwest Tornadoes—2012
Texas Tornadoes—2012
Southeast/Ohio Valley Tornadoes—2012
Southern Plains/Southwest Drought and Heat Wave—2011
Texas, New Mexico, Arizona Wildfires—2011
Tropical Storm Lee—2011
Hurricane Irene—2011
Rockies and Midwest Severe Weather—2011
Missouri River flooding—2011
Midwest/Southeast Tornadoes and Severe Weather—2011
Mississippi River flooding—2011
Midwest/Southeast Tornadoes—2011
Southeast/Ohio Valley/Midwest Tornadoes—2011
Midwest/Southeast Tornadoes—2011
Southeast/Midwest Tornadoes—2011
Midwest/Southeast Tornadoes—2011
Groundhog Day Blizzard—2011
Arizona Severe Weather—2010
Oklahoma, Kansas, and Texas Tornadoes and Severe Weather—2010
East/South Flooding and Severe Weather—2010
Northeast Flooding—2010
Southwest/Great Plains Drought—2009
Western Wildfires—2009
Midwest, South and East Severe Weather—2009
South/Southeast Severe Weather and Tornadoes—2009
Midwest/Southeast Tornadoes—2009

(continued)

Table 7.1 (continued)

Event—Year
Southeast/Ohio Valley Severe Weather—2009
Widespread Drought—2008
U.S. Wildfires—2008
Hurricane Ike—2008
Hurricane Gustav—2008
Hurricane Dolly—2008
Midwest Flooding—2008
Midwest/Mid-Atlantic Severe Weather—2008
Midwest Tornadoes and Severe Weather—2008
Southeast Tornadoes and Severe Weather—2008
Plains/Eastern Drought—2007
Western Wildfires—2007
East/South Severe Weather—2007
Spring Freeze—2007
California Freeze—2007
Numerous Wildfires—2006
Widespread Drought—2006
Northeast Flooding—2006
Midwest/Southeast Tornadoes—2006
Midwest/Ohio Valley Tornadoes—2006
Severe Storms and Tornadoes—2006
Hurricane Wilma—2005
Hurricane Rita—2005
Midwest Drought—2005
Hurricane Katrina—2005
Hurricane Dennis—2005
Hurricane Jeanne—2004
Hurricane Ivan—2004
Hurricane Frances—2004
Hurricane Charley—2004
Severe Storms, Hail, Tornadoes—2004
California Wildfires—2003
Hurricane Isabel—2003
Severe Weather—2003

(continued)

7.14 Pineapple Express

Table 7.1 (continued)

Event—Year
Severe Storms/Tornadoes—2003
Severe Storms/Hail Early—2003
Western Fire Season—2002
Widespread Drought—2002
Severe Storms and Tornadoes—2002
Tropical Storm Allison—2001
Midwest/Ohio Valley Hail and Tornadoes—2001
Western Fire Season—2000
Drought/Heat Wave—2000
Hurricane Floyd—1999
Drought/Heat Wave—1999
OK-KS Tornadoes—1999
AR-TN Tornadoes—1999
Winter Storm—1999
California Freeze—1998
Texas Flooding—1998
Hurricane Georges—1998
Southern Drought/Heat Wave—1998
Hurricane Bonnie—1998
Severe Storms, Tornadoes—1998
Minnesota Severe Storms/Hail—1998
Southeast Severe Weather—1998
Northeast Ice Storm—1998
Northern Plains Flooding—1997
MS and OH Valleys Flood/Tornadoes—1997
West Coast Flooding—1997
Hurricane Fran—1996
Southern Plains Drought—1996
Pacific Northwest Severe Flooding—1996
Blizzard/Floods—1996
Hurricane Opal—1995
Hurricane Marilyn—1995
South Plains Severe Weather—1995
California Flooding—1995

(continued)

Table 7.1 (continued)

Event—Year
Western Fire Season—1994
Texas Flooding—1994
Tropical Storm Alberto—1994
Tornadoes—1994
Southeast Ice Storm—1994
Winter Damage, Cold Wave—1994
California Wildfires—1993
Southeast Drought/Heat Wave—1993
Midwest Flooding—1993
Storm/Blizzard—1993
Nor'easter—1992
Hurricane Iniki—1992
Hurricane Andrew—1992
Severe Storms, Hail—1992
Hail, Tornadoes—1992
Severe Storms—1992
Oakland Firestorm—1991
Hurricane Bob—1991
Severe Storms, Tornadoes—1991
California Freeze—1990
Hail Storm—1990
Southern Flooding—1990
Winter Damage, Cold Wave, Frost—1989
Hurricane Hugo—1989
Northern Plains Drought—1989
Severe Storms—1989
Drought/Heat Wave—1988
Southeast Drought Heat Wave—1988
Hurricane Juan—1988
Hurricane Gloria—1988
Hurricane Elena—1988
Florida Freeze—1985
Winter Damage, Cold Wave—1985
Tornadoes, Severe Storms, Floods—1984

(continued)

Table 7.1 (continued)

Event—Year
Florida Freeze—1983
Hurricane Alicia—1983
Gulf States Storms and Flooding—1983
Western Storms and Flooding—1983
Severe storms—1982
Severe Storms, Flash Floods, Hail, Tornadoes—1981
Drought/Heat Wave—1980
Hurricane Allen—1980

until the next one inevitably happens by 2150. These enormous storms occur about every 200–250 years; old-timers call them *Pineapple Expresses*, because they originate near Hawaii.

The last Big One hit Oregon and California in the winter of 1861–62. A California witness to the largest weather-related disaster to hit the US in the last 200 years described the event as follows:

> The winter of 1861–2 was a hard one. From November until the latter part of March there was a succession of storms and floods. I remember … being in Weaverville, I think it was in the month of December, 1861…. It had been raining all the day previous. The ground was covered with snow one foot deep, and on the mountains much deeper. … It rained all afternoon and night. The weather had turned warm, and the rain came down in torrents…. We slept until about 4 o'clock in the morning, when Jerry Whitmore, one of Uncle Joe's partners, came to where we were, and knocked on the door to wake us up. Uncle Joe called out, 'What is wanted?' Jerry replied, 'The bridge is gone—not a stick left, and the water will soon be up to the house'.
>
> Standing on high ground, one could see property of all kinds on its way to the ocean. The river itself seemed like some mighty uncontrollable monster of destruction broken away from its bonds, rushing uncontrollably on, and everywhere carrying ruin and destruction in its course. … one could see floating down parts of mills, sluice-boxes, miners cabins, water-wheels, hen-coops, parts of bridges, bales of hay, household furniture, sawed lumber, old logs, huge spruce and pine trees that had withstood former storms for hundreds of years—all rushing down that mad stream on their way to the boundless ocean. From the head settlement to the mouth of the Trinity River, for a distance of one hundred and fifty miles, everything was swept to destruction. Not a bridge was left, or a mining-wheel or a sluice-box. Parts of ranches and miners cabins met the same fate.[19]

First-hand witnesses described the Pineapple Express—later called the *Great Flood of 1862*—as weeks of continuous rain beginning in November 1861 and ending 43 days later. The constant downpour filled valleys from the Columbia River on the border between Washington and Oregon, and valleys south to

---

[19] Carr, John (1891).

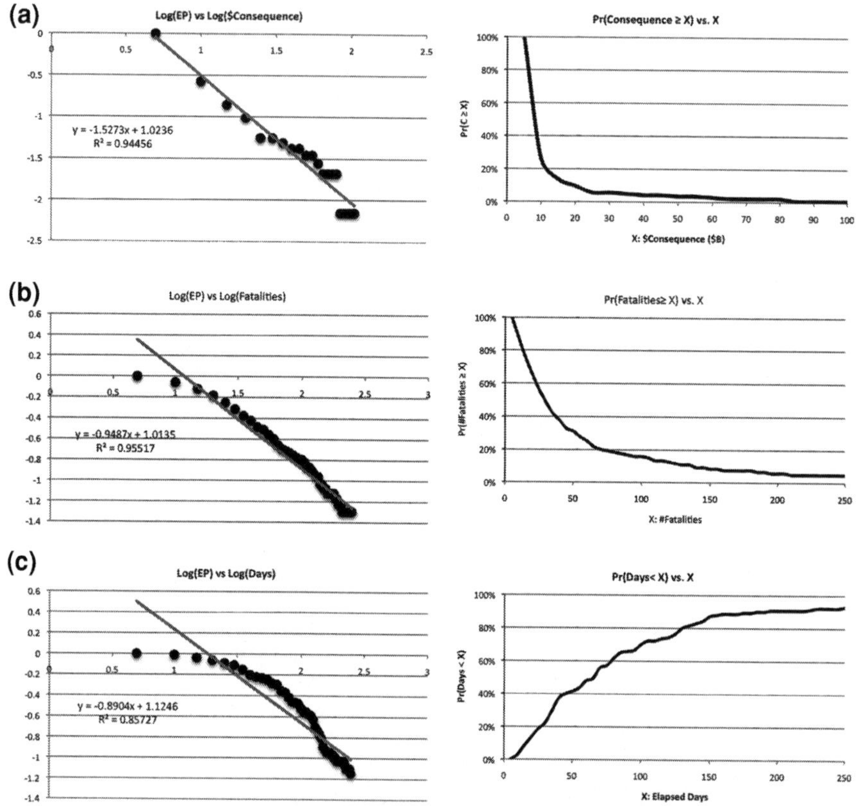

Fig. 7.7 US disasters 1980–2012 in excess of $1 billion obey long-tailed distributions in terms of cost, fatalities, and elapsed time between disasters (Levy walk). **a** Exceedence and probability of cost of disasters. **b** Exceedence and probability of fatalities, and **c** Exceedence and probability of the elapsed time between disasters are long-tailed.

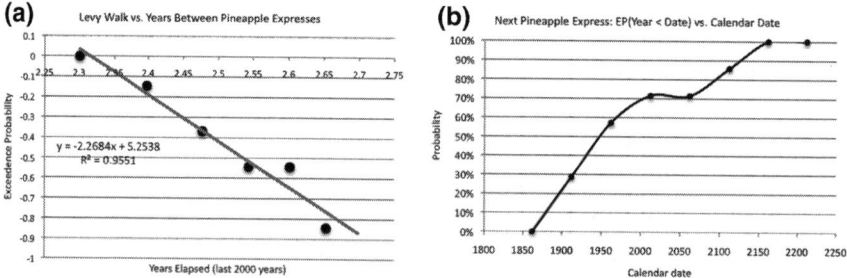

Fig. 7.8 The probability of the next pineapple express is 70 % or greater. The next atmospheric river to flood central California is certain to devastate the state between now and 2150. **a** Elapsed time between successive Pineapple Expresses are Levy walks with a long-tailed distribution. **b** Probability of the next Pineapple Express based on (**a**) forecasts the next event will happen between now and 2150

California including the Central Valley down to San Diego. Then the weather turned warmer, melting snows in higher elevations and sending swollen streams of water into valleys below. So much of the taxable real estate in the state of California was destroyed that the State of California slipped into bankruptcy. State employees and legislators went without pay for a year and a half afterward, and the newly elected governor, Leland Stanford, attended his inauguration in a boat.

The Pineapple Express is an example of an *atmospheric river*. These are "rivers in the sky"—massive jet stream-propelled moisture-laden currents that move through the upper atmosphere from the Pacific Ocean to the West Coast. A USGS exercise called ARkStorm[20] simulated the probable impact of the next major Pineapple Express and the prognostication isn't good.[21] When the next Great Flood hits California with a punch equal to the 1861–62 storm, it will cause $725 billion in damages and affect one-fourth of all homes in California. The protracted downpour of ARkStorm will dump the equivalent of 15 Mississippi rivers into a lake 300 miles long and 20 miles wide. Central Valley will be ten feet under the lake's water for an indefinite amount of time. California's state capital, Sacramento, will lie on the bottom of the inland ocean.

Winds will reach 125 mph, landslides will block roads and destroy houses, and agriculture—California is the world's fifth largest supplier of food and agriculture commodities—will go offline, indefinitely. Power, water, sewer, and emergency services will cease. Businesses will be out of business, and the economy will stop. In total, the damage from ARkStorm will cost three times the damages expected from a severe California earthquake.

Pineapple expresses come in all sizes. Smaller ones like the Hawaiian storm that ravaged San Francisco in 1952, the Christmas flood of 1964, and Willamette Valley (Oregon) flood of 1996 are miniature atmospheric rivers that occasionally rain down on the West Coast, leaving destruction in their wake. These floods are self-similar fractal long-tailed patterns headed towards bigger and more frequent black swans.

Can we afford to play Russian roulette with global warming?

References

Carr, John (1891). "Pioneer days in California". *Times publishing company*, pp. 291–295, 397.
Gregory, J. M., P. M. Foster (2008), Transient climate response estimated from radiative forcing and observed temperature change, J. Geophysical Research, v. 113, D23105, (2008).
Hansen, James, Makiko Sato, Pushker Kharecha, Karina von Schuckmann, *Earth's Energy Imbalance and Implications* (2012a), NASA Goddard Institute for Space Studies, January 2012. New York, NY 10025, USA Columbia University Earth Institute, New York, NY 10027, USA.

---

[20] ARkStorm is a play on words: AR means atmospheric river; k means 1000, and Storm conjures up Noah's Ark and the Biblical flood.
[21] http://en.wikipedia.org/wiki/ARkStorm

Hansen, J., Makiko Sato, and R. Ruedy, 2012b: Perception of climate change. *Proc. Natl. Acad. Sci.*, 109, 14726-14727, E2415-E2423, doi:10.1073/pnas.1205276109

National Academies Press (2008). *Severe Space Weather Events–Understanding Societal and Economic Impacts: A Workshop Report.* http://www.nap.edu/catalog.php?record_id=12507

Rebecca Lindsey (August 3, 2010), "What if global warming isn't as severe as predicted?" http://earthobservatory.nasa.gov/blogs/climateqa/what-if-global-warming-isnt-as-severe-as-predicted/

Singer, Fred, and Dennis T. Avery (2005), *The Physical Evidence of Earth's Unstoppable 1,500-Year Climate Cycle.* S. Fred Singer President, Science and Environmental Policy Project, Adjunct Scholar National Center for Policy Analysis, and Dennis T. Avery, Senior Fellow, Hudson Institute. NCPA Policy Report No. 279, September 2005 ISBN #1-56808-149-9.

Trenberth, K. E., J. T. Fasullo, J. Kiehl (2009), Earth's Global Energy Budget, *American Meterological Society*, March 2009, pp. 311–323.

Waymer, Jim (2010), "Study finds link between sun and hurricanes", http://usatoday30.usatoday.com/weather/storms/hurricanes/2010-06-01-hurricanes-sun_N.htm

Zhen-Shan, L. and S. Xian (2007), Multi-scale analysis of global temperature changes and trend of a drop in temperature in the next 20 years. *Meteorology and Atmospheric Physics*, 95, 115–121, 2007.

Bombs

8

Abstract

Can the next big event be predicted? It is possible to detect the next bomb before it goes off? World events are not random—they may be long-tailed, but not entirely unpredictable. Take population as an example. It is exploding. And it is on the move: from East and South to North and West. Global migration follows wealth, freedom, and opportunity. More over, massive migratory patterns foretell an impending tipping point—a nonlinear consequence of a tilted world. As the North and West gets richer and its labor force diminishes, it also edges closer to a socio-economic tipping point. And, as the East and South becomes overpopulated with working age youth, it becomes hungrier for consumer goods like automobiles, cellular telephones, vacations, and fashionable clothes. This tension is accelerated by the "natural" tendency for wealth to accumulate according to Pareto's long-tailed distribution. And to make things even more stressful, ownership of most wealth is concentrated in a handful of mega-corporations—largely financial organizations—that rule the economic world from within the shadows of highly interconnected financial webs. The imbalance between the "have's and have not's" is widening, because of preferential attachment—the simplest form of self-organization.

8.1 The Malthusian Catastrophe

In 1968 young and impressionable men were volunteering for vasectomies because the hero of the era, Paul Ehrlich (1932), Stanford University Professor of Population Studies and author of the wildly popular book, *Population Bomb*, "warned of [impending] mass starvation of humans in the 1970s and 1980s due to overpopulation."[1] Better to stop reproducing now, than risk starvation! Ehrlich wasn't the first person to extrapolate human fecundity and conclude that we are all doomed unless we change our sex habits. Fortunately, his prediction was a spectacular failure. Instead of 7 billion starving denizens, today's 7.5 billion inhabitants of earth are doing very nicely, indeed.

Like Ehrlich, Reverend Thomas Malthus (1766–1834) predicted an inevitable *Malthusian catastrophe* in his 1798 bestseller, *An Essay on the Principle of Population*. Europe had recovered from the Black Plague to "go forth and multiply" its surviving inventory of Homo sapiens through religiously motivated reproduction. But Reverend Malthus reasoned that unbridled promiscuity could go no further after reaching a tipping point. Both Ehrlich and Malthus argued that resources like food and water expanded arithmetically while humans multiplied geometrically. If food production doubled, the number of hungry mouths would quadruple. The math was simple—sooner or later, population would outstrip resources and the *paradox of enrichment* bubble would burst.

The population of earth in 1798 was one billion; by 1968 it was 3 billion; and now it exceeds 7 billion. Global population is projected to top 11 billion by 2050, assuming human fertility levels off as wealth increases—the other *"wealth effect"*. Prognosticators of nearly every age have proclaimed the end of humanity, but it never happens. Why not? What happened to the Malthusian catastrophes predicted in 1798 and again in 1968? Why were two world-class thinkers so wrong?

Malthus's dire prediction was made on the eve of one of the most dramatic leaps in human history. He overlooked a technological punctuation of black swan proportions—the *Industrial Revolution* to be exact. Industrialization made agriculture radically more productive through mechanization, which meant more people could be fed at lower cost. It transformed agriculture just as radically as it transformed manufacturing, growth of cities, and explosive population growth. Instead of ending the middle age's growth spurt, industrialization accelerated growth even more.

Similarly, Ehrlich proclaimed the impending demise of earthlings on the eve of another dramatic leap—the *Green Revolution*. Norman Borlaug (1914–2009), the "Father of the Green Revolution … credited with saving over a billion people from starvation, [through] development of high-yielding varieties of cereal grains, expansion of irrigation infrastructure, modernization of management techniques, distribution of hybridized seeds, synthetic fertilizers, and pesticides to farmers" ruined Ehrlich's prediction[2] Borlaug's Green Revolution saved more lives than perhaps any other invention in history. Borlaug received the Nobel Peace Prize, the Presidential Medal of Freedom, and the Congressional Gold Medal, and yet few people living today even know his name.

---

[1] http://en.wikipedia.org/wiki/The_Population_Bomb
[2] http://en.wikipedia.org/wiki/Green_Revolution

It seems the most notable people make the worst predictions right before the most earthshaking punctuations. Conversely, the least known innovators seem to make the really important and profound earthshaking leaps right when we least expect them. One moment we are all going to die, and the next moment we are all thriving and growing. Innovative breakthroughs seem to happen on the eve of destruction.

Fortunately for humanity, Malthus and Ehrlich were dead wrong. Neither catastrophe befell humanity. Or were they simply too early?

8.2 Old, Poor, Crowded, and Oppressed

It is highly probable that the earth's ecosystem can support 11 billion people without a Malthusian catastrophe. The population bomb may not explode because of sheer population numbers. Instead, the ticking time bomb is likely to go off in places where humans are old, poor, crowded, and oppressed. This demographic bomb could set off a chain of connected events leading to the next global calamity. The challenge facing humanity is to mitigate this black swan before it is too late. But, what is the challenge and what exactly is the innovation needed on the eve of destruction this time?

The number of international migrants is at historic highs, estimated to affect more than 200 million people, annually.[3,4] Approximately 3 % of the world's population is on the move at any point in time, according to United Nations' Global Migration Database. And the size of this migrant stock is rising. [Essentially, this equates with the entire population of the world moving every 30 years, or the entire population of the US moving every 18 months].

Most migrants voluntarily move to another region or country. But not always. The United Nations' High Commission for Refugees says more than 43 million people were forcibly displaced in 2010. Pakistan received the most refugees (1.7 million people) followed by Iran (1.1 million), and Syria (one million). The statistics are troublesome:

1. Forty percent of the globe's refugees are in Africa.
2. Conflict (war and social unrest) is the leading cause of refugees (27.1 million in 2009).
3. In 2009, refugees came from Afghanistan (2.9 million people), Iraq (1.7 million), Somalia (678,300), Democratic Republic of Congo (455,900), and Burma (406,700).
4. Refugees (60 %) seek out developed countries with a high standard of living: In 2010, Europe had the highest number of migrants (70 million), followed

---

[3] Migration, Agriculture and Rural Development: A FAO Perspective, Eight Coordination Meeting on International Migration, New York 16–17 November 2009.

[4] United Nations, Department of Economic and Social Affairs, Population Division (2009). International Migration, 2009 Wallchart (United Nations publication E.09.XIII.8).

by Asia (61 million), and North America (50 million). The United States hosts nearly one in five international migrants in the world.
5. More than 20 % of the workforce of North America and 10 % of the workforce of Europe comes from migrants.
6. The "gateways to freedom" are found along borders: US-Mexican, Chinese-Russian, and European-Russian.
7. Migrants are older than the average age of the global population, suggesting that populations in developed countries are getting older for two reasons: older people are moving to developed countries, and developed country's fertility rates are low, which means they need labor. 36 % of the world population is under 20 years old while 15 % are migrants.

Old migrants leave poor, crowded, oppressed countries for richer, less-crowded, freer countries where other old people already live. For example, people are leaving Afghanistan, Angola, Benin, Burkina Faso, Burundi, Haiti, Madagascar, Mali, Niger, Republic of the Congo, Somalia, Uganda, Yemen, Zambia, Zimbabwe, and moving to the UK, Germany, France, Canada, United States, Australia, and other prosperous, free, and aging countries. In 2010, migrants made up only 1.4 % of the population in Latin America; 1.9 % in Africa; 1.5 % in Asia, but 14.2 % in North America. People are moving from South to North and East to West.

Why? Generally, the world's population is shifted by *preferential attachment*: migrants follow wealth. In 2010, wealth per adult was highly concentrated in Europe and North America. Specifically, 32 % of global wealth was in Europe, 31 % in North America, and 24 % in Asia/Oceania. Conversely, 4 % was in Latin America, and only 1 % in Africa. No wonder people are moving from the South to the North and East to West.

This movement makes the populations of Europe and North America even older. And older populations are less productive; more likely to burden social welfare programs; and less likely to advocate for change. Europe and North America will continue to experience labor shortages and immigration challenges, and Africa/Asia will continue to experience labor surpluses and employment challenges. These imbalances can only lead to social and political unrest. One might conclude that the world will eventually tilt too far and fall over, if this form of self-organization does not reverse. What can be done about this mega-trend?

8.3 The Wealth Effect

Economists define the *wealth effect* as the propensity of people to spend more if they are rich or think they are rich. But the real effect of wealth on the world's population is to precipitously reduce fecundity.[5] Rich people have fewer children.

---

[5] Fertility, or the ability/desire to reproduce.

8.3 The Wealth Effect

Table 8.1 Shifting workforce demographics throughout the world

Region	2010		2050	
	15–64 (workforce) (%)	≥65 (retired) (%)	15–64 (workforce) (%)	≥65 (retired) (%)
Latin America	8.5	6.9	7.9	19.5
North America	5.2	13.1	4.7	22.0
Europe	11.1	16.3	6.8	27.4
Africa	12.9	3.4	22.3	7.1
Asia (Saudi peninsula)	61.8	6.7	57.8	17.3
Oceania (Australia)	0.5	10.9	0.5	18.7
World		7.6		16.2

The United Nations' Department of Economic and Social Affairs, Population Division, predicted in 2004 that world population would stall at around 9.22 billion by 2075, because of the wealth effect. This projection is based on the idea that fertility will drop below replacement rates as prosperity increases. This drop in reproduction reduces the size of local workforces, and in turn, the number of people providing financial support for aging populations. By 2100, Europe's working population will be cut in half, from 12 to 5.9 %, while Africa's workforce will double—from 13 to 25 %. Without immigration, Europe will have too many old people to support and Africa will have too many young people to employ.

Table 8.1 illustrates the dramatic shift in workforce and retirement-aged demographics for the five major regions of the world. Within the next 35 years the number of retired people will double according to the United Nations.[6] Simultaneously, the workforce will concentrate in Africa and Asia and bring with it even greater pressure on immigration. If the trend of increased migration from young, poor, crowded, and oppressed regions to old, rich, relatively un-crowded, and politically free regions continues, Europe and North America must prepare for an onslaught of immigrants.

But a much bigger problem exists.

8.4 The World Is Tilted

Thomas Friedman (1953–) made his fortune declaring the "world is flat", but the flat world is also tilted. Nearly two-thirds of the global workforce lives in Asia (China, India, Middle East), while nearly two-thirds of the capital lives in North America and Europe. That is, flatland is tilted towards Asia when it comes to labor and tilted towards the West when it comes to capital. If the law of supply and

---

[6] http://hdr.undp.org/en/statistics/data/mobility/map/Share of world's population.

demand actually exists in global economics, either capital must flow to Asia, or labor must flow to the West. [This partially explains the migratory trend of people moving from South to North and East to West].

In the late 20th century capital flowed towards labor in Asia, but as emerging nation's labor becomes more expensive, capital flows will reach a point of diminishing returns. Of course, cheap labor in Africa and South America is waiting to take the place of China and India. As long as there are several billion people willing to work for less than $1.00/h, capital will find cheap labor. Will this trend continue into the 21st century?

Clearly, the *comparative advantage* of cheap labor will continue to work in favor of emerging nations for some time. But it won't last forever as the wealth effect improves living standards and resources become limited. Eventually, China and India will become like the West—and their people will decide to go shopping. When this tipping point is reached, labor and capital flows will reverse. Asia is headed for a consumer economy, while the West is headed for even greater labor shortages and immigration challenges.

What is the nature of this mega-shift in capital and people? Two themes stand above all others—the rapid emergence of *distributed production* concurrent with continued *concentration of capital*—and the global control that capital concentration implies. As capital and labor adjust to the imbalance between cheap labor and the West's command over capital unfolds over the next century, production will become less concentrated in the West while capital becomes more concentrated. The Western nations are getting richer by distributing manufacturing throughout the world, placing it closer to customers.

8.5 Distributed Manufacturing

One of the central concerns of governments throughout the world is the viability of their country's manufacturing industries. The strongest nations have the strongest manufacturing industrial bases. As labor costs equalize, transportation and resource efficiencies begin to dominate, making local manufacturing more attractive (vs. outsourcing to poorer countries). Instead of highly centralized manufacturing plants located in one or two countries, the 21st century industrial titans will be forced to distribute manufacturing across the globe. Products are more likely to be manufactured by the people that consume them.

For example, the largest automobile manufacturer in the world, Toyota, produces parts and assembles trucks and autos in 26 countries: Argentina, Australia, Belgium, Brazil, Canada, China, Colombia, Czech Republic, France, Ghana, Indonesia, Japan, Mexico, Pakistan, Philippines, Portugal, Russia, South Africa, Thailand, Turkey, United Kingdom, United States, India, Vietnam, and Zimbabwe. GM Chairman and Chief Executive Officer Daniel F. Akerson said in February 2011, "seven out of 10 of our vehicles were made outside the United States" and, "we have 11 joint ventures" with government-owned Chinese companies.[7] Similarly, German Volkswagen Group had 61 factories in fifteen European countries and six countries in the Americas, Asia, and Africa.

Boeing Commercial Airplane Company was heavily criticized for outsourcing much of the 787 *Dreamliner*, but the truth is that distributed manufacturing made the Dreamliner possible. Not only were costs kept in check ($32 billion), but marketing benefitted by spreading the manufacturing wealth to potential customers around the world: wings were built in Japan; horizontal stabilizers in Italy; fuselage sections in South Korea, Italy, USA; passenger doors in France; cargo doors, access doors, and crew escape door in Sweden; software development in India; floor beams in India; electrical wiring in France; landing gear in UK and France; and air conditioning packs in USA. Final assembly is done in Everett, Washington.[8]

Manufacturing powerhouses like Boeing impress politicians, because they create jobs. The Everett assembly plant was visited by, "former U.S. President Bill Clinton; former U.S. Vice President Al Gore; former Texas Governor and U.S. President George W. Bush; former Russian President Boris Yeltsin; Chinese President Jiang Zemin; Prime Minister Paul Keating of Australia; Prime Minister Mahathir bin Mohamad of Malaysia; President Ion Iliescu of Romania; Prince Philippe of Spain; President Meri of Estonia; the late King Hussein of Jordan; His Royal Highness Prince Andrew, The Duke of York; President Megawati Sukarnoputri of Indonesia; former U.S. Speaker of the House of Representatives Dennis Hastert; former U.S. astronaut Neil Armstrong; and Crown Prince Shaikh Salman bin Hamad Al-Khalifa of Bahrain."[9] But the reality is that most of the airplane was made outside of the US.

8.6 Concentrated Financial Control

Manufacturing may be spreading across the globe but control of capital is not. In a study done by researchers at ETH-Zürich—Eidgenössische Technische Hochschule, Zurich, Switzerland, the previously hidden network structure of transnational corporations was found to be dominated by a handful of large, connected, wealthy financial institutions.[10] Senior researcher Stefano Battiston, and colleagues James Glattfelder and Stefania Vitali mapped out the network of 43,060 interconnected transnational corporations taken from the ORBIS 2007 database.[11] They discovered a surprising structure.

---

[7] http://www.factcheck.org/2012/06/is-gm-becoming-china-motors/
[8] http://en.wikipedia.org/wiki/Boeing_787_Dreamliner
[9] http://en.wikipedia.org/wiki/Boeing_787_Dreamliner
[10] Vitali et al. (2011).
[11] ORBIS is a proprietary database containing information on millions of companies produced by Bureau van Dijk. http://www.library.hbs.edu/go/orbis.html.

The ETH network represented corporations as nodes and ownership as weighted links—stocks owned by investors. If company A owns 51 % of the stock of company B, then node A is connected to node B by a weighted link labeled 0.51. Additionally, majority ownership implies 100 % ownership. Company A controls company B, because 51 % is a majority position. The resulting network is a web of ownership and power among the global corporations.

Indirect control of company A over C is exerted via a chain of links from company A to B to C. A *strongly connected component* (SCC) is defined as a subset of the network in which every node exercises control over every other node through a chain of indirect control. That is, a handful of nodes (companies) form a core of interconnected companies that are indivisible but exert undue influence over all other nodes. The researchers discovered that a mere 737 top holders had accumulated 80 % of the control over the value of all transnational corporations on the planet!

Moreover, a large proportion of the strongly connected component core is made up of banks of one form or the other. Capital formation and investment decisions are largely under the control of a *scale-free network* with a relatively small SCC core.[12] "Nearly 4/10 of the control over the economic value of transnational corporations in the world is held, via a complicated web of ownership relations, by a group of 147 transnational corporations in the core, which has almost full control of itself," say Vitali, Glattfelder, and Battiston.

8.7 The World Is a Bow Tie

Transnational corporations have had centuries to evolve and form scale-free networks, because preferential attachment is always at work and 100–200 years is a long time. So it is not surprising that the ETH network is dominated by a handful of highly connected and highly influential companies. The average degree (number of connections) of the ETH network core is 20, which means the network is relatively sparse. Most of the top companies in the core are well known:

Barclays PLC
AXA
State Street Corp
JPMorgan Chase
Legal and General Group
UBS AG
Merrill Lynch
Deutsche Bank
Franklin Resources
Credit Suisse Group
Natixis

---

[12] Boss et al. (2004).

8.7 The World Is a Bow Tie

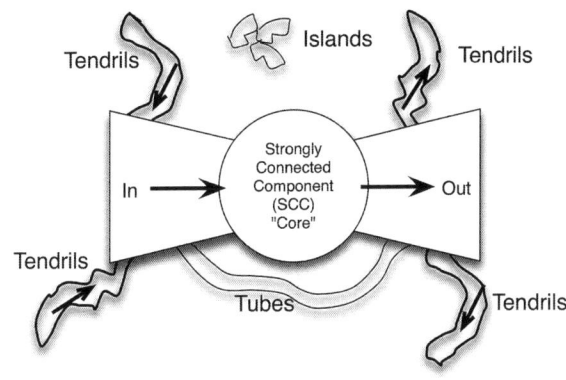

Fig. 8.1 The massive self-organized bow tie structure of networks typically contain a strongly connected component in the middle, large numbers of input and output nodes and tendrils, one or more tubes that bypass the core, and a few isolated islands. Bow tie structures have been observed in global corporation networks and the Internet

Goldman Sachs
T. Rowe Price
Legg Mason
Morgan Stanley
Mitsubishi UFJ Financial Group
Northern Trust Corp
Societe Generale
Bank of America
Lloyds TSB Group
INVESCO PLC
Old Mutual Public Ltd
Aviva PLC
Schroder's PLC
Lehman Brothers
Sun Life Financial
Standard Life PLC
CNCE
Nomura Holdings

Note that Lehman Brothers was a prominent member of the top holders in the SCC core. The ETH network was constructed in 2007—before the financial crash of 2008. The fact that Lehman Brothers also crashed in the 2008 financial meltdown supports the claim that core members exert massive influence over global capital. This connectivity supports the "financial contagion" theory—highly connected networks are vulnerable to collapse when a hub node fails.

The ETH study shows that capital formation and control residing in the SCC core is part of a highly evolved bow tie—a scale free network containing a secondary structure shaped like a bow tie, see Fig. 8.1. There are many minor companies on the periphery—one group provides inputs and another group provides outputs, but little else. Then a series of tendrils feed into the IN group and fan outward from the OUT group. In addition, one or more tubes bypass the

SCC—these tubes contain a chain of influential corporations that ignore the powerful core. Finally, a very small number of isolated "island nodes" coexist with the dominant core. They have very little influence.

The bow tie model has been observed before—it is called the *World Wide Web*. Broder et al. studied the Internet structure consisting of over 200 million servers and found that the massive Internet was scale-free exactly like the bow tie structure shown in Fig. 8.1.[13] But the World Wide Web is less evolved than the transnational corporation network. Its core is a strongly connected component just like the transnational corporation network, but much bigger and less connected. The WWW core consists of 56 million (27 %) of all servers known in 1999, and 17 million (8.3 %) disconnected islands.

The global cyberspace structure of society in the 21st century is mimicking the global physical space structure of capitalism that emerged 200 years after the Industrial Revolution exploded onto the scene. Apparently, big networks tend to all look the same—a bow tie. Why?

8.8 Concentration of Power Is a Network Effect

Does the fact that global financial control concentrates in a relatively small number of corporations mean there is a conspiracy to dominate the world? Is greed the reason? No. Instead, the bow tie structure of both networks—transnational corporations and the World Wide Web—is a natural consequence of *network science*, not greed or conspiracy. To prove this claim, consider the following computer simulation.

The SCC emergence simulator represents corporations as nodes and ownership links in a network model similar to the one explored by the ETH scientists. Starting with 500 nodes and zero links the computer algorithm percolates the network by randomly selecting a node and linking it to another node selected according to its value. That is, the likelihood of linking a randomly selected node to another node is proportional to the value of the node. The value of a node is equal to the corporation's revenues in Fig. 8.2a, and the degree (number of connections) in Fig. 8.2b. Value-preferential attachment seeks high-revenue ownership, while degree-preferential attachment seeks high popularity ownership.

The computer simulator constructs a network from the 2011 Global 500 corporations ranked according to their revenues. The top ten G500 corporations are listed in Table 8.2. Starting with zero links, the simulation simply repeatedly applies the following preferential attachment algorithm: link a randomly selected G500 corporation to another G500 corporation with probability proportional to its revenues. Do the same for another 500-node network, using degree-preference instead of value-preference, and compare the results. Are they different?

---

[13] Broder et al. (2000).

8.8 Concentration of Power Is a Network Effect

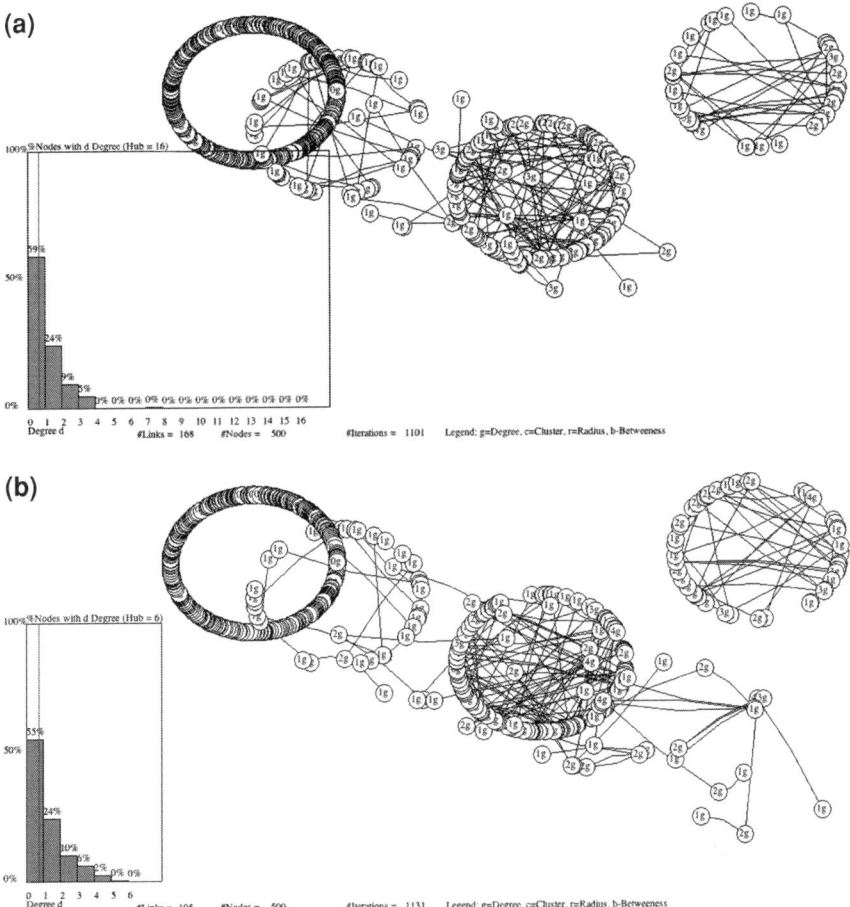

Fig. 8.2 Strongly connected components naturally emerge from value-preferential and degree-preferential attachment in large highly evolved networks. The resulting scale-free G500 networks in **a** and **b** with strongly connected components of approximately 18 % of the total produced by random collapses (5 % of the nodes) at random times (12 % of the time) are nearly identical regardless of value or degree attachments. **a** Value-preference attachment: the largest strongly connected component contains 18 % of the G500 nodes (shown in the *center*), and is a scale-free network with long-tailed connectivity (shown in the bar chart). **b** Degree-preference attachment: the largest strongly connected component contains 18 % of the G500 nodes (shown in the *center*), and is a scale-free network with long-tailed connectivity (shown in the *bar chart*)

Intuition says these two networks should evolve into two different types of networks, because one is based on the revenues of the G500 and the other is based on the popularity of connected nodes. Revenues and popularity are apples and oranges, so why should the networks be similar? Does intuition serve us well, here?

Table 8.2 Revenues of the top 10 Global 500 Corporations in 2011 exceeded $2.8 trillion

G500 company	2011 revenues ($M)
Wal-Mart Stores	$421,849
Royal Dutch Shell	$378,152
Exxon Mobil	$354,674
BP	$308,928
Sinopec Group	$273,422
China National Petroleum	$240,192
State grid	$226,294
Toyota Motor	$221,760
Japan Post Holdings	$203,958
Chevron	$196,337
	$2,825,566

Revenues of all 500 global companies are extreme—they obey a long-tailed distribution with exponent of 1.5889

If left to its own emergent devices, a large strongly connected component will emerge from the G500 simulation, regardless of the metric used to determine connectivity. Every company will eventually be included. The resulting network will be scale-free with one node dominating all others. No bow tie structure, however! Without some form of intervention, preferential attachment leads to a single monopoly hub.

But this scenario is not representative of the real world. Instead, business networks are occasionally attacked by *black swan* events such as a slumping economy, poor management, war, etc. So the computer model must be adjusted to reflect the reality of business and the economy. The emergence simulation must accommodate occasional failures, just like the real world.

Suppose an occasional disaster is allowed to strike and dismantle parts of the network. That is, the emerging network is randomly attacked. The attack partially dismantles the network by removing one or more nodes (corporations that fail) and replaces them with new nodes (creative destruction). New nodes with revenues similar to G500 companies are inserted into the evolving G500 network.

The emergence simulation takes out N nodes with probability P, every year of the simulation. Most years nothing happens, but one node is replaced with a new node with probability P to simulate *creative destruction*. Parameters N and P were chosen so that the largest connected component in the emergent network approximates the results obtained by the ETH scientists (P = 18 %).

Figure 8.2 shows what happens after hundreds of exchanges in the simulation of G500 network formation. Regardless of the objective—preference for value or degree—the results are nearly identical! Both algorithms produce scale-free networks. Both produce a small core of strongly connected nodes. Both networks emerge because of network effects, alone. The resulting *structure* has nothing to do with greed, power, control, or superior performance of one company over another. Rather, the structure of the ETH network and the simulated networks shown here are identical, simply because preferential attachment produces identical structure.

Global corporations and nodes of the World Wide Web are governed by natural network effects that arise solely due to mathematics. Self-organization through preferential attachment and an occasional collapse determines the size and structure of the network. The organizing principle—revenues or number of connections—does not matter as much as self-organization.

But, simulation cannot predict which company will rise to dominate all others. This is a matter of randomness. Similarly, self-organization cannot determine which person or country will dominate all others, even though economic inequality is built into preferential attachment. Only a few actors are allowed to rise to the top.

8.9 Pareto Was Right of Course

Two-thirds of global wealth is controlled by approximately 10 % of the population located mainly in Europe and North America, while two-thirds of the workforce lives elsewhere. This tilt in distribution of wealth has always been with us, according to Vilfredo Pareto, the Italian political economist who first observed it. Vifredo Federico Damaso Pareto (1848–1923) showed that in every past and present culture, regardless of political structure, a very small minority of the population owns a very large portion of everything worth anything. That is, wealth inequality is like a scale-free bow tie. Regardless of governmental programs to equalize wealth across economic classes, wealth distribution obeys a long-tailed distribution—the *Pareto distribution*.[14]

Pareto was born in Paris, studied at the University of Turin, lived in both Italy and Switzerland, and married a Russian woman. He was a citizen of Europe. The Fascist government of Italy promoted his ideas to justify Mussolini's rise to power in the 1920s. Mussolini made him Senator of the Kingdom of Italy in 1923, but Pareto never served. Nonetheless, Mussolini used wealth inequality to incite voters in his rise to power in pre-war Italy.

Pareto was an engineer, sociologist, economist, political scientist, and philosopher—not a politician. He may have been the first *big data analytics* scientist of the modern era, having collected and studied tax records, rental incomes, and personal income records from Germany, Britain, Prussia, Saxony, Ireland, Italy, and Peru. Regardless of the century, country, or political structure of a nation, the distribution of wealth was always the same—80 % of all wealth was owned by 20 % of the people. Pareto's law is sometimes called the, "80–20 % law." The *Pareto Index* is equivalent to the fractal dimension of the long-tailed Pareto Curve.

---

[14] The Pareto distribution is a power law.

The 80–20 % rule is a long-tailed power law like the others studied in this book. Instead of following a bell-shaped normal distribution like Pareto expected, wealth across all ages and societies follows a long-tailed power law distribution. "It is a social law in the nature of man", according to Pareto. Today, the Pareto index is a measure of the inequality of income distribution. Another Italian, Corrado Gini (1884–1965), a theoretical fascist ideologue wrote *The Scientific Basis of Fascism* in 1927, which carried the Pareto banner even further. The Gini index is a measure of financial inequality used today. [One of the major tools of capitalism was invented by a Fascist socialist].

The Gini index G is related to the fractal dimension q as follows:

$$G = \frac{1-q}{q+1}; \quad 0 \leq q < 1; \quad G = \frac{q-1}{q+1}; \quad q \geq 1$$

When G = 0, wealth is distributed "fairly", according to Gini. The Gini index for the US, for example, hovers around G = 0.47, which equates with a fractal dimension q = 2.8. Gini index values for the UK and Germany steadily increased during the 1990s from 0.55 and 0.42, to 0.70 and 0.60, respectively.[15] Thus, wealth is more evenly distributed in the US than wealthy European nations where social programs tend to be more comprehensive.

Fractal dimension increases with inequality, which means the long-tailed distribution becomes shorter—not longer. Contrary to catastrophe theory described previously, longer-tailed distribution equals more equality. Short-tailed distribution equals less equality. That is, fractal dimensions less than 1.0 indicate greater equality. For example, a long-tailed Pareto with q = 0.75 equates to a Gini index of G = 0.14.

Attempts to flatten the Pareto distribution have repeatedly failed. Governments have tried various forms of wealth redistribution without much success. While taxing the rich to give to the poor, increasing taxes, decreasing taxes, welfare programs, and shifts in various laws mitigate the inequality for a short time, they ultimately fail to flatten Pareto's distribution. During economic hard times, poor people get poorer, and during good economic times, rich people get richer. Why?

8.10 Econophysics: Wealth Is a Gas

There must be something fundamental about Pareto distributions because they keep recurring throughout history and governments cannot defeat them! What is it? Answer: wealth distribution is governed by physics. That is, wealth accumulation can be modeled as a physical property of the economy just like Brownian motion in a room full of gas molecules can be modeled as colliding balls. *Econophysics* is a new branch of science that applies physics to economic processes. It explains Pareto wealth concentration, or what physicists call

---

[15] Clementi and Gallegati (2005).

condensation. Like moving gas molecules, wealth drifts and diffuses its way through its container.

Jean-Philippe Bouchard and Marc Mezard combined finance and physics in 2000 to answer the question left unanswered by Pareto: "why does wealth fall on a long-tailed distribution curve?"[16] Bouchaud and Mezard claim that the Pareto distribution emerges out of a random process with one simple organizing principle:

Wealth increase is proportional to the amount of wealth already owned.

Furthermore, wealth accumulates in a small fraction of the population because random wealth exchanges between pairs of individuals in the population favor the wealthier of buyer and seller, in amounts proportional to the wealth difference between buyer and seller. In other words:

$Increase_i = growth\_rate (Wealth_i - Wealth_j)$
Where growth_rate is assumed constant and $Wealth_i > Wealth_j$ are the wealth of individuals i and j, selected at random from a population of n individuals.

This model is so simple it is unlikely to work! To illustrate that it does indeed explain Pareto's long-tailed distribution, suppose a computer is used to repeat this rule thousands of times on an initial population of people with nearly identical—but slightly different—wealth. Over time, a disproportionate transfer of wealth will take place—from less wealthy people to slightly wealthier people, simply because of preferential attachment. That is, a small difference in wealth will magnify as interactions between buyers and sellers repeatedly transfer wealth from one person to the other according to the rule above. Ultimately, all wealth will "condense" or concentrate with one or two individuals!

Figure 8.3a illustrates the result of simulating wealth transfer as described above. After 50,000 exchanges in a population of 200 individuals with nearly equal—but slightly different—initial wealth, a long-tailed distribution appears. The distribution is long-tailed because one individual (at the extreme right—100 %), is far wealthier than all others. It is a power law, because the log-log plot inserted in the upper right portion of the graph falls on a straight line. Finally, it shows the extreme end state of repeated applications of wealth transfer, because most everyone (83 %) are relatively poor, while a very small group of elites are middle class, rich, or even super-rich.

According to the econophysics theory of wealth accumulation, wealth drifts and diffuses throughout a population as described above. But, instead of spreading out, the simulation described here condenses wealth. That is, starting from a long-tailed or flat distribution, wealth decreases among the less fortunate, causing the long-tailed distribution to drift to the left or low end. At the same time, wealthier

---

[16] Bouchaud and Mezard (2000).

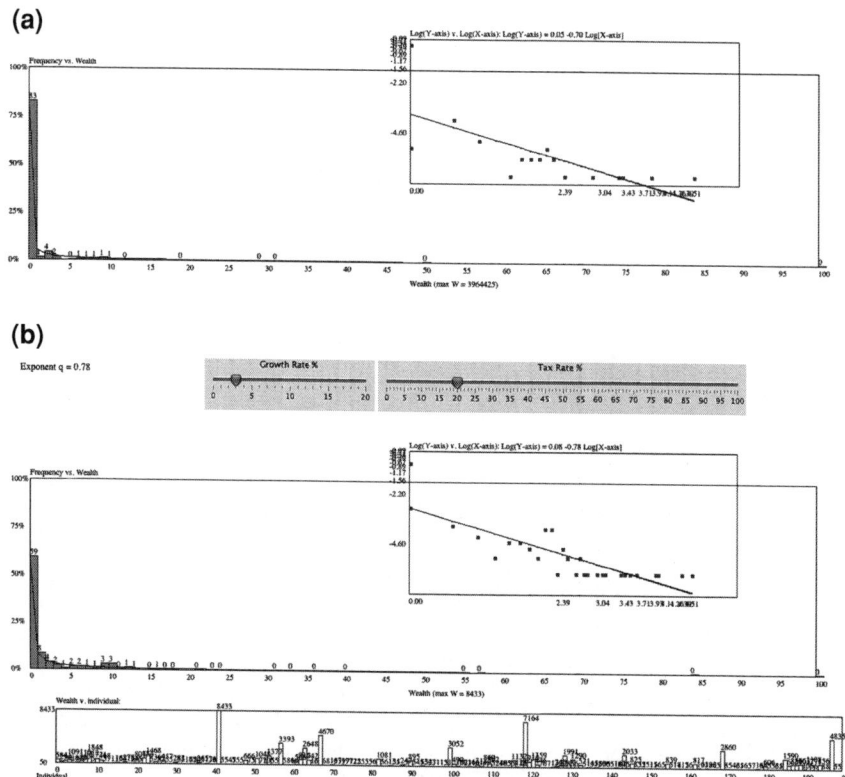

Fig. 8.3 Pareto distributions with and without taxes: (a). No taxes or redistribution. (b). Wealth increases are taxed at 20 % and the proceeds are periodically redistributed to individuals with less than average wealth

individuals become rarer as they drift to the right or high end. Overall, the Pareto curve becomes shorter, rather than longer. In physical terms, the diffusing gas gathers in one corner of the container, instead of spreading throughout.

8.11 Redistribution

What can be done about wealth inequality? Suppose taxes are levied to redistribute wealth to individuals of below-average wealth. Each transaction between buyer and seller is taxed and the accumulated taxes are equally distributed back to below-average individuals. The result of simulating this is shown in Fig. 8.3b, assuming a 20 % tax. Note that the distribution is still long-tailed, but there is a bigger middle class. Most individuals fall below average, while a few rise above average. But a small elite core still remains far more prosperous than the average.

8.11 Redistribution

This naturally raises the next question, "Is it possible to flatten the Pareto curve by redistributing wealth?" To answer this question, I ran a series of simulations that assumed a growth rate (profit) of 3 % on each transaction, and varied the tax rate applied to everyone to see taxation's effect. Taxes of 20–80 % were collected and redistributed to the less-than-average-wealthy individuals. The exchange transactions were repeated as many as 150,000 times per simulation. Initially, every individual (n = 200) is given a random amount of wealth ranging from zero to $10. Pairs of individuals were selected at random and 3 % of the wealth difference between the two is transferred to the wealthiest individual. The computer repeats this thousands times.

For example, after 50,000 exchanges, total wealth accumulated rose to $3,641,620 ($18,208 per capita).[17] The power law fractal dimension was 0.82 and 63 % of the population landed on the bottom of the economic scale (far left of the power law). When taxes were raised to 50 %, 55 % of the population landed at the bottom, but per capita wealth was only $254 after 50,000 exchanges. Per capita wealth rose to $132,865 after 100,000 exchanges, with 87 % of the population falling to the bottom (the long-tailed distribution fractal dimension is 0.69).

When taxes were increased even further, say 80 %, per capita wealth remained very low—$16 after 50,000 exchanges, $45 after 100,000 exchanges, and $314 after 150,000 exchanges. At the other extreme, zero taxes applied to the same 200 individuals and 3 % growth, netted over $1,652,440 per capita after 50,000 iterations! The no-taxes scenario landed 85 % of the population at the bottom of the economic scale, and a Pareto distribution with fractal dimension of 0.76. These numbers are illustrative, only.

8.12 Wealth Redistribution Isn't Natural

Regardless of the level of taxation, a Pareto distribution emerges over and over again from the econophysics simulation. These results imply that chance, rather than investment ability, is the main source of long-tailed wealth distribution. The Pareto distribution cannot be defeated, but it can be bent. That is, we can change the fractal dimension of the distribution by redistribution of wealth through taxes and other socio-political means, but we cannot completely flatten it. But, there is a price to be paid for attempting to level the curve.

A *low-tax strategy* leads to rapid rise in overall prosperity, leaving over 50 % of the population at the lowest level of the economy. Transfers never reach equilibrium until one or two percent of the population owns everything. Low taxes promote rapid economic growth and a handful of super wealthy people at the top. Typical fractal dimensions of 0.75 or lower are characteristic of low-taxes (the 80–20 % rule equals a fractal dimension of 1.1). Recall that the lower the fractal

---

[17] An exchange is one buyer-seller transaction.

dimension, the longer the tail of the Pareto curve, and the more equally distributed is wealth.

A *high-tax strategy* leads to stagnation or a very slow accumulation of national wealth, also leaving over 50 % or more at the lowest level of the Pareto curve. But the Pareto distribution is much shorter-tailed, with typical fractal dimensions of 1.2 or more (larger fractal dimensions mean shorter tails). Unlike the low-tax scenario, high taxes eventually lead to stability (stagnation) whereby additional transactions no longer enrich the population or individuals.

Both strategies produce Pareto distributions, but the high-tax strategy produces a shorter-tailed distribution with lower overall prosperity. In other words, high taxes leads to an overall lower standard of living for everyone. The low-tax strategy produces higher overall prosperity, but greater inequality! Low taxes accelerate wealth accumulation—especially for the rich. This suggests a counter-intuitive result:

> Overall wealth of an economic system increases with inequality and decreases with equality. Equality equals poverty. Inequality equals prosperity.

A third strategy should be considered: progressive taxation whereby tax rates increase with wealth accumulation. A gradual increase in taxation after a country becomes wealthy tends to distribute wealth without significantly reducing overall prosperity. Wealth distribution still follows a long-tailed Pareto curve. But inevitably, rising taxes—even for the wealthy—eventually leads to stagnation of the economy. When this happens, an occasional reduction in taxes is necessary to stimulate further economic advancement.

References

Bouchaud, J.; Mezard, M. (2000). "Wealth condensation in a simple model of economy". Physica A: Statistical Mechanics and its Applications 282 (3–4): 536. Bibcode:2000PhyA..282..536B. doi:10.1016/S0378-4371(00)00205-3.

Broder, A., R. Kumar, F. Maghoul, P. Raghavan, S. Rajagopalan, R. Stata, A. Tomkins, and J. Wiener (2000), Graph structure in the web, Computer Networks, 33, 309 (2000).

Boss M, Elsinger H, Summer M, Thurner S (2004), Network topology of the interbank market. Quantitative Finance 4: 677–684.

Clementi F., and M. Gallegati (2005), "Pareto's Law of Income Distribution: Evidence for Germany, the United Kingdom, and the United States," Papers physics/0504217, arXiv.org, revised Mar 2006.

Vitali S, Glattfelder JB, Battiston S (2011) The Network of Global Corporate Control. PLoS ONE 6(10): e25995. doi:10.1371/journal.pone.0025995.

Leaps

9

Abstract

Success in the 21st century depends on making leaps—sizeable Levy Flights into the future. Cautious steps won't do, because they are too timid and too late. Projects like the 100YSS (100-year Star Ship), hunt for the X-Prize, and filling new white spaces will be the norm rather than the exception. Globalization increases competition and competition increases the rate of change. Everyone will operate at the long-tail end of the exceedence probability distribution. The innovation Levy Flight curve must become shorter. How can these Levy Flights take place more often and how can their impact be magnified? Governments and corporations will play an increasingly important role, but they must change their business models to achieve even greater leaps forward. In addition, the author offers two themes for the individual: un-constrain the problems and look for unfilled white spaces.

9.1 Not Waiting for ET

Mae Jemison was born in Alabama but raised in Chicago. She is an extreme human. She also epitomizes the 21st century—a rare period of human history where leaps are the norm. She was the first black woman to travel into space in 1992. She holds nine honorary doctorates in science, engineering, letters, and the humanities. And in 2012 she became the face of the 100-year star ship program created to leapfrog humanity into Interstellar space travel. Like I said, Mae Jemison is an extreme human.

Jemison's resume is full of leaps: she received a B.S. in chemical engineering and a B.A. in African and African–American Studies from Stanford University in 1977, and then a medical degree from Cornell University Medical College in 1981. She served in the Peace Corps from 1985 to 1987. Her first ride into space in 1992 was followed by a TV appearance in an episode of *Star Trek: The Next Generation*. When asked why she quit her medical practice to go to work for NASA, she said it was easier to be a shuttle astronaut, "than waiting around in a cornfield, waiting for ET to pick me up or something."[1]

Jemison founded the *Dorothy Jemison Foundation for Excellence* in honor of her mother. Its goal is to promote science of both types—physical and social. In 2012 her foundation was awarded $500,000 by the *Defense Advanced Research Projects Agency* (DARPA) to form the independent, non-governmental, long-term *100 Year Starship* (100YSS) program. Its task is to make human interstellar flight possible in the next 100 years. One small step for woman—one big leap for mankind.

DARPA's 100YSS program aims to develop persistent, long-term, private-sector investment in the science of long-distance space travel. Like the *US Space Program* before it, the purpose of 100YSS is to kick-start the research and development needed by the Department of Defense, NASA, and the private and commercial sector to colonize other solar systems.

The challenge is extreme. Traveling at the speed of the Space Shuttle, it would take 165,000 years to reach the nearest star system, Alpha Centauri, 4.3 light years away, or 25, 277,600,000,000 miles! A space ship capable of accelerating to the speed of light could reduce travel time to 5 years. But so far, nobody knows how to accelerate mass to the speed of light. Can the time be reduced to less than 100 years—a constant speed of 8,000 miles per second? The fastest space craft ever built by humans orbited the sun in 1974 and 1976. Helios II reached speeds of 50 miles per second—less than 1 % of the speed required to reach Alpha Centauri in less than 100 years. Can anyone survive a trip outside of the solar system that takes so long? That is the big question.

Mae Jemison and the 100YSS initiative characterize the 21st century. Gone are the corporate R&D labs of the 20th century. Gone are shortsighted government programs to pave roads, subsidize solar energy, or stop global climate change. Gone is incremental thinking and visionary journeys of a thousand small steps. The 21st century is about waves, surges, bubbles, and leaps. Innovation alone will not be enough in this century. Nothing less than leaps are required.

9.2 Levy Flights of Fancy

Invention, innovation, and breakthroughs in science and technology are long-tailed Levy flights in time as explored in *Bak's Sand Pile*. The impact of such innovations in dollars or social value is much more difficult to estimate, but I speculate that the monetary or social impact is also a long tailed phenomenon. Invention and innovation comes in waves, not smooth increments. For example, the exceedence probability distribution for "Internet inventions" shown in Fig. 9.1 illustrates the

---

[1] Peterson (2004).

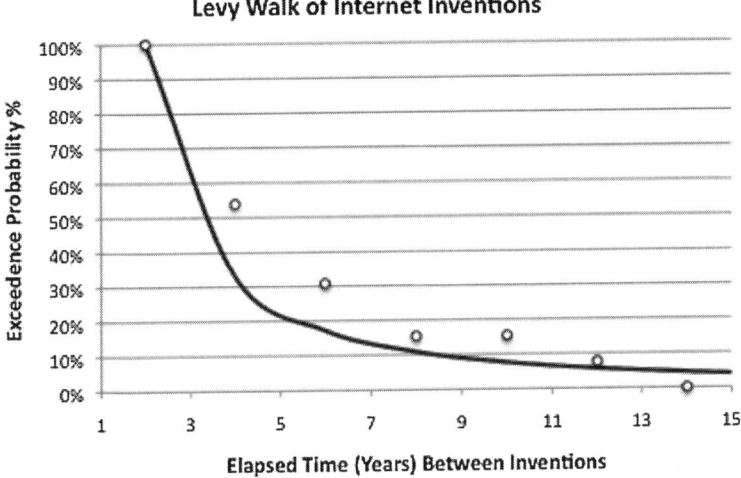

Fig. 9.1 Elapsed time between Internet inventions and innovations follows a long tailed distribution. In the future, this distribution must become shorter to keep pace

long-tailed nature of leaps in technology. Progress is bursty and punctuated by rare leaps forward.

Given that the human race depends on ample innovation to support its rapid expansion, it is the goal of many nations to extend the long tail of impact and shorten the long tail of elapsed time between breakthroughs. We want more breakthroughs as significant as the Internet to happen sooner rather than later. The goal is to move ahead in leaps and bounds, rather than at a snail's pace. For this to happen, the "significance distribution" has to be long-tailed, while the "elapsed time distribution" has to be short-tailed. We need major breakthroughs to happen with much higher probability, separated in time by shorter periods.

How does humanity shorten the long tail of innovation's Levy Flight in time? During the 20th century capital formation played a key role in progress. During the 21st century the trick is to leverage capital formation and focus it on leaps instead of slow burns. The 21st century won't just happen—instead it has to be manufactured. If we want to reach Alpha Centauri in less than 100 years, we have to change the game of invention and innovation. How?

9.3 X-Prize Leaps

Amir Ansari is an inventor and serial entrepreneur from Tehran, Iran who immigrated to the US before the revolution sent that country back into the dark ages. Educated in electrical and computer engineering at George Mason University, the immigrant entrepreneur made his fortune by building innovative Internet switches and selling them to Fortune 500 companies such as WorldCom, Qwest Communications, Hewlett-Packard, Sprint, and AT&T.

In 2004 Ansari joined the *X-Prize Foundation's* Board of Trustees, and together with his family, sponsored the first *X-Prize*. The idea of rewarding invention and innovation goes back to the $25,000 Orteig Prize offered in 1919 by French hotelier Raymond Orteig for the first nonstop flight between New York City and Paris. Charles Lindbergh won the prize by flying the Spirit of St. Louis across the Atlantic in 1927. Ansari followed in Orteig's footsteps.

Interestingly, nine aeronautical teams spent $400,000 in pursuit of the $25,000 Orteig Prize. The idea of spending $400,000 to win $25,000 in prize money impressed entrepreneur Peter Diamandis, co-author of *Abundance: The Future Is Better Than You Think*, and founder of the *X-Prize Foundation*.[2] Diamandis gained global attention when he announced a $10 million prize to the first privately financed team that could build and fly a three-passenger vehicle 100 kilometers into space—twice—within two weeks. This became the *Ansari X-Prize for Suborbital Spaceflight*. And like the Orteig Prize before it, the Ansari X-Prize motivated 26 teams from seven nations to spend more than $100 million to win the $10 million prize.

On October 4, 2004, *Mojave Aerospace Ventures* won the Ansari X-Prize in a Burt Rutan designed spacecraft called *SpaceShipOne*. Microsoft co-founder and billionaire Paul Allen and legendary aircraft designer Burt Rutan founded the company. Also in 2004, the company partnered with Richard Branson's *Virgin Galactic* to develop the *Virgin SpaceShip*—a commercial craft designed to transport tourists into suborbital bliss. Mojave Aerospace Ventures is in it to make history while also making money.

The nonprofit X-Prize Foundation's mission is to focus capital on radical breakthroughs for the benefit of humanity. The foundation calls this "*incentivized competition.*" It takes capital formation beyond venture capital and corporate R&D. It inspires formation of new industries and markets that are stuck due to insurmountable failures or a commonly held belief that a solution is not feasible or possible. It motivates and inspires brilliant innovators from all disciplines to leverage their intellectual and financial capital. And so far, it has worked extremely well.

If you want a glimpse of the next 20 years look into the X-Prize Foundation's prizes as of 2012:

> The Progressive Insurance Automotive X-Prize ($10 million) for creating a clean car that gets 100 MPGe—miles-per-gallon-electric car equivalent.
> The Archon Genomics X-Prize ($10 million) will be awarded to the first group to sequence 100 human genomes within 10 days or less.
> The Google Lunar X-Prize ($30 million) for the first team to put a robot on the moon, wander around, and send pictures back to Earth.
> The Northrop Grumman Lunar Lander X CHALLENGE ($2 million) is a competition to build precise, efficient small rocket systems.

---

[2] Diamandis and Kotler (2012).

The Wendy Schmidt Oil Cleanup X CHALLENGE ($1 million) was won in 2011 by the *Chabot Space and Science Center* team for their innovative solution for cleaning up marine oil spills from tankers, ocean platforms, and other sources.

The Qualcomm Tricorder X-Prize ($10 million) to create a portable medical diagnostic device like the tricorder seen on the TV series StarTrek.

The Nokia Sensing X CHALLENGE ($2.25 million) to stimulate the development of a new generation of health sensors and sensing technologies that can drastically improve the quality, accuracy and ease of monitoring a person's health.

9.4 Leaps of Faith

The idea of *incentivized competition* spread like a plague to one of the most risk-averse corners of the world—the US government. In 2009 the government offered a combined total of $44.5 million of taxpayer-supplied incentive to any and all takers through the Department of Defense, Department of Energy, Environmental Protection Agency, NASA, and Health and Human Services, according to a report by the Congressional Research Service.[3] The innovations ranged from a better light bulb to combat vehicles that find their own way around a battlefield.

For example, Department of Defense allocated millions in prize money for leaps of progress in wearable power packs, autonomous combat vehicles; Department of Energy funded the Freedom Prize (reduce dependence on oil), Hydrogen Prize (hydrogen storage and power), Bright Tomorrow (low-power lighting), Progressive Automotive Prize (100 mpg + cars); EPA opened the American Le Mans Series Green challenge Race (green transportation), NASA funded the Astronaut Glove Challenge (improved space suit gloves), and robotics for exploring other planets.

But, these leaps are still not extreme enough, because the problems we face are too wicked. They defy traditional methods. To become really innovative, we have to change how complex problems are addressed. This will require new ways of thinking.

9.5 Un-constrain Your Constraints

Mae Jemison and the X-Prize Foundation innovators leaped over mere mortals and solved seemingly unsolvable problems by hacking their assumptions and removing problem constraints. Jemison nullified the assumption that a black woman could not travel into space, and the X-Prizes leap over entire governments and enable a handful of talented people to do what most countries cannot do. They illustrate a trait of the 21st century—how to solve wicked problems by un-constraining them.

---

[3] Stine (2009).

According to Wikipedia, a *wicked problem* "is a problem that is difficult or impossible to solve because of incomplete, contradictory, and changing requirements that are often difficult to recognize."[4] Solutions are difficult because of the inherent complexity of the problem. For example, border security and immigration policy have been persistent long-term problems for the US because they involve many connections, dependencies, controversies, stakeholders, and contradictions. Policies regarding the legalization of marijuana, abortion, counter-terrorism, and other persistent and complex issues fall into this category as well.

Charles Churchman (1913–2004), an American philosopher and professor of Business Administration at the University of California, Berkeley coined the phrase in 1967.[5] But the colorful management science phrase has more star power. Actually, wicked problems are very easy to understand, because they are simple complex problems. That is, they are simple to understand, but difficult to solve.

Wicked problems have been around for a long time. Mathematicians defined them in precise terms more than a century ago as a class of problems that take *combinatorial* effort to solve as the size of the problem slowly grows. For example, a puzzle with 6 pieces takes 720 steps to sort out and arrange into a picture, and another puzzle with 12 pieces takes 479,001,600 steps.[6] Doubling the size of the problem from 6 pieces to 12 increases the effort by a factor of 665,280—a combinatorial amount. Simple puzzles are complex or wicked problems to solve, because of *combinatorial explosion*—the rapid rise in possible solutions for marginal increases in problem size.

Regardless of terminology, it should be apparent to the reader by now that most of the easy problems plaguing humanity have been solved. Only the wicked ones remain. Therefore, future policy wonks, politicians, scientists, technologist, humanists, ecologists, and leaders in general need an approach equal to the task. Moreover, most of us are not mathematicians, so a toolbox of simple mental exercises is called for. One such tool is called *unconstrained optimization*. In simple terms, it means:

> When confronted by a wicked or complex problem that defies solution, remove constraints, introduce measured randomness or chaos, and let the unconstrained system find a new optimal (fixed) point.

For example, a computer algorithm that *randomly* pieces together a puzzle with 12 pieces is faster than an algorithm that methodically tries all possible combinations. Who would have thought that a random approach to piecing together a puzzle would be quicker than a well-thought-out process? In general, puzzles can be "solved" much faster by random trial-and-error than by exhaustive enumeration. Many real-world problems are quickly solved by computerized chance. *Monte Carlo* techniques were used to build the first Atomic bomb back in the

---

[4] http://en.wikipedia.org/wiki/Wicked_problem

[5] Churchman (1967).

[6] 6! versus 12!.

9.5 Un-constrain Your Constraints

1940s and these techniques continue to be used to solve more mundane business problems, today. But there is more.

Unconstrained optimization is based on a simple assumption: solutions to complex problems tend to seek stable fixed points (configurations), because stability increases survivability or resilience. When destabilized by a threat, war, new technology, disaster, disruption, or change, every complex system seeks a new state of relative tranquility. If the system succeeds, it survives. If not, it is eliminated. Complex systems tend to self-organize such that disorder is replaced by order, and disarray is replaced by structure. The trick is to disrupt your own assumptions and let a certain amount of chaos enter the process.

The new stable point typically optimizes some quantity like energy, cost, effort, etc. A disrupted complex system may become chaotic or temporarily out of control, but eventually it seeks and finds a new stable configuration—or it goes extinct. The process of finding stability is called *emergence*, and emergence typically works from the bottom up to the top. That is, small local changes combine and reinforce each other, spreading to mid level structures, and eventually to high levels.

Obvious examples of unconstrained optimization finding new fixed points are the unprecedented and unexpected rise of the Internet and online social networks (Twitter and Facebook), movements in the stock market (booms and busts), collaborative filtering online (recommended books by Amazon.com), structure of cities (both physical and virtual as in *The Sims*), legal and political systems (forms of government), ecosystems (both biological and industrial), *innovation butterflies* (sudden domination of the cell phone market by Apple Inc), emergence of leadership in groups, and the emergence of life itself on Earth.

9.6 Like Prometheus Stealing Fire

Y-HP. Percival Zhang grew up in Wuhan, China, but through a series of fortuitous moves, ended up in the United States as a world-class biochemist working for Virginia Tech.[7] After studying biochemical engineering at East China University of Science and Technology in Shanghai, he became interested in a complex wicked problem—how to transform the petroleum-based economy of the modern world into an environmentally clean and abundantly cheap hydrogen economy. "… When I decided to go to the United States to study for my Ph.D., I began to think how my education and interests could address other critical needs. An article in *Science* magazine by Lee Lynd at Dartmouth about renewable energy changed my research focus."

The idea of hydrogen-powered trucks, busses, and cars has been around for decades. Automobile manufacturers have built prototypes, and some cities have even deployed fleets of hydrogen busses. When combined with a fuel cell and

---

[7] http://www.vtnews.vt.edu/articles/

electric motor, hydrogen delivers clean and powerful performance. It is an ideal fuel for transportation, except for two major problems. First, it is difficult to store and transport to where it is needed, and second, it takes more energy to separate hydrogen from water than the energy delivered by "burning" the hydrogen in a fuel cell. For these reasons, the hydrogen future is stalled, and will remain stalled until solutions can be found.

The hydrogen economy is stuck in low gear because it mimics the petroleum economy. In place of oil refineries, hydrogen refineries along the coast would separate seawater into hydrogen and oxygen, send it through pipes or via tanker trucks to service stations where the highly volatile hydrogen would be stored until pumped into cars and trucks. Then, vehicles consume the hydrogen much like they currently consume gasoline. This model is inefficient for several reasons—separating hydrogen from water takes more energy than you get back from burning hydrogen, and the infrastructure for storing, transporting, and consuming hydrogen in cars and trucks is dangerous and expensive.

Zhang joined Lynd at Dartmouth and received his Ph.D. in 2002 under the famous biochemist. Their article about "enzymatic hydrolysis of cellulose" was the most-cited paper in Biotechnology and Bioengineering in 2004 and was among the most downloaded papers seven years in a row from 2004 to 2010.[8]

By 2005 Zhang was ready to start his own career and work on a complex problem of global significance. Zhang approached the problem of producing, storing, and transporting hydrogen energy by un-constraining the problem, allowing for a period of random trial-and-error exploration to occur, and then cutting to the heart of the question. Zhang asked, "Why not develop an enzyme cocktail to convert sugars directly into hydrogen to power a fuel cell, and why not make the reaction occur directly onboard a vehicle?" If successful, this would be like Prometheus stealing fire from the gods of energy.

Percival Zhang went around the problem by redefining it. He removed the constraints of separating hydrogen from water in an energy intensive refinery—and introduced an unpredictable new idea—use plant and animal enzymes to do the separating with little or no added energy. As for storage—simply ignore it. Instead, make the whole system compact enough to replace the internal combustion engine onboard a truck, bus, or car. His idea is to manufacture hydrogen where it is consumed.

In Zhang's vision, automobiles run on sugar water and enzymes! Zhang's cocktail flows from an onboard tank through a fuel cell that converts it into electricity, which in turn, powers an electric motor. A small battery is needed to initially heat the cocktail to 90° so the hydrogen separation reaction can begin and the driver need not wait for the cocktail to ferment. Every truck, bus, and car becomes a refinery, storage, and delivery system on wheels.

Zhang's sugar water cocktail is a type of disruptive rootkit.

---

[8] Zhang and Lynd (2004).

9.7 Disruptive Rootkit

According to Wikipedia a *rootkit* is, "a stealthy type of software, often malicious, designed to hide the existence of certain processes or programs from normal methods of detection and enable continued privileged access to a computer.[9] Rootkits are stealthy and powerful, because they operate at the most privileged level within a system. The term comes from the Unix operating system, which controls most of the Internet. But more recently, it has been applied more generally. Rootkits are stealthy mechanisms for getting past bureaucratic roadblocks and constraints so that things can get done. It is a strategy for making progress against all odds.

The all-time spectacular rootkit of the early 21st century was the introduction, by Apple Inc., of the iPad tablet. The iPad disrupted the Wintel duopoly held for two decades by Microsoft and Intel. It was so upsetting that Dell Computer was forced into the hands of private owners, Hewlett-Packard lost its top spot as the global leader of personal computer sales, and Apple became the largest market-cap company listed in the New York Stock Exchange.

The Apple Inc tablet struck at the roots of the PC industry. Apple's iOS, a derivative of Unix, obsolesced the non-Unix Windows operating system. Moreover, iOS obsolesced Apple's own operating systems! The iPad defined a new touch interface, and brought an even larger segment of consumers into computing by making it possible for anyone to browse the Internet, photograph friends, communicate by videophone, and download software applications easier and at an order of magnitude cheaper than ever before. Apple's rootkit redefined computing for the masses and re-wrote industry rules.

The disruptive iPad illustrates how changing the rules can disrupt an entire industry at its foundation. It illustrates how to leap ahead by hacking your rootkit and making your own rules.

9.8 Fill a White Space

Tamara Carlton runs the Innovation Leadership Board in association with the Stanford University Foresight and Innovation program in Palo Alto, California. In addition to leading ILB, Tamara serves as a Forum for Innovation Fellow at the U.S. Chamber of Commerce Foundation, exploring themes of innovation and abundance that affect American growth and prosperity long-term. She specializes in helping people find and fill white spaces.[10] The phrase *white space* is used to describe the unused gap in bandwidth in electronics, but more recently has become a descriptive term for an under-utilized, under-recognized, or under-rated business

---

[9] http://en.wikipedia.org/wiki/Rootkit
[10] Aka white spots.

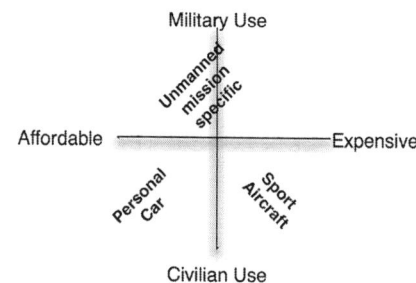

Fig. 9.2 Carlton's white spot analysis illustrates how to identify niches and opportunities in the gaps between established industries and/or product lines. A two-dimensional analysis of the flying car white spots identifies at least three opportunities

niche. White spaces are market niches where a company might have an advantage in a crowded playing field.

Innovators look for white space opportunities. But how? According to Carlton's Playbook for Strategic Foresight and Innovation, "... [the white spot] method provides a quick way to organize competing solutions and similar examples against two dimensions in order to help you assess areas of promising opportunity and search for high growth markets."[11] Carlton's white spot method places examples of competing businesses along two dimensions such as a cost dimension (high-cost vs. low-cost), and an application dimension (consumer vs. military uses).

Tamara describes Kevin, an aeronautics engineering graduate student at Stanford University, who had always enjoyed stories of flying cars. He was now in a position to propose the concept for real funding, and he wanted to use the white spot method to help him show possible partners and investors why the timing was right for flying cars. See Fig. 9.2.

After reviewing earlier notes, he identified cost and utility as his two most critical dimensions for an opportunity analysis. Cost became the basis for his x-axis, and utility became his y-axis. Kevin then defined opposing endpoints for each axis. While multiple spectrums could exist for either axis, such as high/low and easy/hard, he ultimately selected affordable and expensive as end points for cost and then military and civilian as end points for utility.

With the two axes defined, he now faced an empty 2 × 2 matrix. Kevin's next task was to see if he could place any examples in the various quadrants, particularly at the four extreme corners of the matrix. Kevin started by recalling his earlier jobs at Boeing and NASA, as well as drawing on recent research into the topic. He also spoke with several industry experts to add more examples on the matrix. After plotting all these examples on the matrix, he eventually found an open space—the "white spot"—which marked a potential market opportunity for personal flying cars.

---

[11] www.innovation.io/playbook, p. 127.

White spot analysis is another way to relax constraints on thinking and allow for random thinking to draw you into broader solution spaces. It opens up problem solving to an examples-driven, somewhat trial-and-error process, but with discipline. Change the dimensions on the problem and it leads to new insights. Change the cost dimension to commute time, and the flying car fills a different niche—perhaps reducing commuter time by leaping over freeway traffic.

9.9 Work in the Weeds

Weeds grow in the cracks between sidewalks and orderly rows of corn, tomatoes, strawberries, and grape vines. Most of us try to remove them, but the innovator looking to leap over groupthink seeks to cultivate weeds. Why? Simply because they grow in the seams between established disciplines and markets. Weeds are untapped markets overlooked by others. Opportunity lurks in between established platforms, industrial sectors, and communities.

Salinas Valley, California, is a simple example that moves my agricultural narrative from metaphor to reality. It is called the *salad bowl of America*, because farmers of the Salinas Valley produce 85 % of all lettuce—and most of the broccoli, spinach, strawberries, and artichokes consumed in the USA. It is a $3.8 billion agricultural powerhouse dramatized by John Steinbeck in books and movies like The Grapes of Wrath and East of Eden.

The second largest source of economic prosperity is tourism ($2 billion), and the third source is provided by an array of educational institutions such as the Defense Language Institute, Monterey Institute of International Studies, California State University at Monterey Bay, Monterey Law School, and the Naval Postgraduate School. Anyone in the world in need of a telephone call interpreter is likely to be connected to one of 200 foreign language-speaking experts from Language Line Solutions, a unique company located in the Monterey-Salinas region.

Here is the problem. There is no white space left in Salinas Valley for growing more lettuce, bringing in more tourists, or educating more college students and military officers. The established sources of prosperity have peaked. More seriously, these industries are declining, prompting city fathers to commission a number of economic development studies. Economic Development Director Jeffrey H. Weir proposed the latest attempt to grow the economy in 2009.[12] His proposal was to:

1. Retain and expand existing businesses.
2. Create a computer database on existing sales.
3. Encourage retail expansion.
4. Explore growing crops as a source of alternative energy.
5. Become a telecommuting "bedroom community" for Silicon Valley.
6. Create a business incubator.
7. Market the area to potential employers.

---

[12] Economic development strategic vision and recommended Action Plan Fiscal Year 2009–2010.

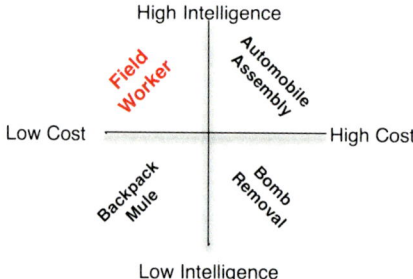

Fig. 9.3 White space analysis using two-dimensional analysis exposes a market niche for robotics in agriculture. An opportunity exists in a highly intelligent machine that can pick lettuce, strawberries, and tomatoes, but it also has to be inexpensive. Similar technology already exists in the application of robotics to automobile assembly and bomb removal by military and law enforcement forces. If simple robots can be developed to carry supplies (backpack mules), an additional white space may open up for low-intelligence and inexpensive consumer robots

This unimaginative list has had little effect simply because it focuses on already tapped-out industries. If production of lettuce increases from 80 to 90 %, so what? The impact will be negligible. Retailers will only expand if the base economy expands. High-tech startups will move to Silicon Valley once they become successful, and business incubators thrive on innovation, not government incentives. The future of Salinas Valley growth is in weeds, not lettuce. Salinas Valley will continue to stagnate until its leaders leap forward by getting down into the weeds.

The solution is very simple: look in the cracks between established industries for opportunity. See Fig. 9.3. Some areas where the Salinas Valley might excel are briefly listed below. The imaginative reader can expand on this short list by focusing on possible intersections between disparate disciplines as I have here.

- Robots and Lettuce: Harvesting crops with robots. Salinas Valley has a labor shortage—few people want to pick strawberries or harvest heads of lettuce now days. Silicon Valley robotic companies are thriving just 50 miles north of the salad bowl. Weeds are literally growing between San Jose and Salinas. Bridge the gap between Silicon Valley robotics and Salinas Valley agriculture and an entirely new industry is born. Figure 9.3 uses Tamara Carlton's white space matrix to visualize this opportunity.
- Internet and Ag: Online produce markets with global reach. Salinas Valley holds an enviable position in the agricultural food chain, but there is no more land available to grow more lettuce or grapes. On the other hand, farmers in Peru, Chile, and around the world have plenty of land, but limited markets. An online market for lettuce NOT grown in Salinas could double the salad bowl's global reach and control of produce. Robust electronic supply chains are thriving in the global electronics industry. Why not build one for produce?
- Wind and NIMBY: Power-generating windmills have lost favor with the public because of NIMBY—Not In My Back Yard. But farmers don't own land for its aesthetic value. Land is a farmer's factory. It can be used to grow lettuce or for

planting windmills. A 500 kw windmill can be a cash crop. 40 such windmills can pay the rent and property taxes as well as provide badly needed power to California's overtaxed grid.[13]

Focusing on the weeds growing in between disparate disciplines often taxes an organization's cultural heritage. It goes against the grain. Farmers know nothing about robots or windmills, for instance, and building out an Internet e-commerce business requires an agricultural background that few Silicon Valley programmers have. Sometimes it is necessary to hack your own culture to get anything done.

9.10 Hack Your Culture

Perhaps the most spectacular failure in 21st century business is the case of Eastman Kodak Company of Rochester, New York. The film giant filed for protection under Chap. 11 on Jan 19, 2012 and was still reorganizing in 2013 as this was written. [It reappeared as a printing company in 2014]. Jonathan Good estimated that 85 billion analog photos were being taken in 2000, around the globe.[14] And most of them were of babies! Kodak derived most of its revenue from printing these photos using their famous silver halide process. One might say that Kodak owned a silver mine.

Just one decade later, nearly the same number of photos was posted on Facebook.com, alone. An additional 300 billion snapshots were taken by billions of people using billions of cell phones. By 2011, Facebook contained 140 billion photos—nearly double the number of analog photos taken in 2000. Three websites: Facebook, Flickr, and Instagram dwarf the Library of Congress in terms of stored photos. People of Earth now take 10 times as many photos today as Eastman Kodak processed the year they went bankrupt! One might ask, "Why did the leading photography company go out of business when the number of photos being taken was expanding exponentially?"

Incredibly, Eastman Kodak not only invented the best chemical process for printing photos, but the giant also invented (and patented) the digital camera back in the 1970s. Kodak dominated the movie industry with its motion picture film, and for a short time, made inroads into medical applications of photography such as X-ray imagery and CT-scanning. In spite of this overwhelming leadership, the king of photography put itself out of business.

Kodak illustrates the punishment a company takes when it ignores the Innovator's Dilemma. But why did the industry giant ignore the obvious shift from analog to digital? In 2001 corporate leaders attended Clayton Christensen's executive leadership course on Innovators Dilemma.[15] Indeed, Kodak was the poster child for

---

[13] Potentially $1 million/year at 19 cents/kwh.

[14] http://blog.1000memories.com/94-number-of-photos-ever-taken-digital-and-analog-in-shoebox

[15] Harvard Business School Professor Clayton Christensen is the world's foremost authority on disruptive innovation and author of *The Innovator's Dilemma: When New Technologies Cause Great Firms to Fail*, Harper Business, 1997.

Christensen's theory. Superficially, the company crashed because it could not transition from a chemical company that made its fortune on selling atoms to a digital imaging company that made money from selling bits. According to Christensen's theory, digital photos replaced silver halide photos because digital is cheaper, faster, and more appealing to consumers. But for Kodak, switching from atoms to bits undermined their *rootkit*—making money from processing film and selling chemicals. Digital photography was considered inferior because of early low quality of digital imaging and the short shelf life of pictures processed on a consumer's printer.

But the real reason Kodak collapsed goes deeper: its culture prevented it from thinking outside of its box. Culture determines what an organization can think and do. And Kodak's culture was built around chemical film processing. The very idea of an image without chemicals was unthinkable. Kodak could not envision any other self-image. Management at the company circa 2000 was horrified by the idea of putting an imager on a cell phone, or making digital movies. Rather than embracing the white spaces and unconstrained optimization that digital image processing opened up, the company hunkered down and made even better film and the cameras to go with it. The company filed for bankruptcy when it had the best chemical photography in its history.

The only way out for Kodak would have been to hack its own culture. That is, management had to kill its culture and replace it with a contrarian one. Step one would require the business model to shift from chemical film to image science—a technology that existed within Kodak Labs. In fact, Kodak Labs was far ahead of everyone else in the industry. Kodak should have harvested the chemical film business and put all of its resources behind digital imaging. That means software development for image sharing; cell phone image transmission, two-way video on tablets, and anything else that used a digital imager. In fact, Kodak's strategy might have been to give away the digital imager for two cents per minute for every minute used by AT&T or Verizon customers who began downloading images by the billions.

Hacking your own culture is another way to unconstraint the problem and inject randomness into an established processes to optimize on an alternative fixed point. It can be unsettling, but it can also save a giant from extinction. The consequence of sticking to Kodak's 130-year-old culture was to lose its entire business. Gone was the culture and gone was the company. Like Salinas Valley lettuce, Kodak dominated photography so thoroughly that little headroom remained in the established market. Better film had limited potential for growth. But better bits had unlimited growth. Unfortunately, the company could not escape the innovator's dilemma because Kodak culture made working in the weeds unthinkable.

9.11 Extreme Exceptions

Mae Jemison, the X-Prize winners, and dozens of other innovators have learned the best way to solve wicked problems is to redefine them. When a problem is so difficult and complex that it seems unsolvable, the innovator steps back and reconsiders: Is the problem overly constrained? Is it formulated right? Is there a sub-problem that is similar, but already solved? Am I so deep into the box that I cannot see out? Am I a prisoner to my own culture?

Leaping will become commonplace in the 21st century, simply because we have no choice—the problems we face are too wicked and demanding to be handled any other way.

References

Churchman, C. W. (1967), "Guest Editorial" of Management Science (Vol. 14, No. 4, December 1967).

Diamandis, Peter H., and Steven Kotler (2012), Abundance: The Future Is Better Than You Think, Free Press, 2012, 400 pp., ISBN-10: 1451614217, ISBN-13: 978-1451614213.

Peterson, Charles A. (2004), "Neward Advocate! Astronaut talks to DU freshmen", September 2, 2004. http://en.wikipedia.org/wiki/Mae_Jemison

Stine, Deborah D. (2009), Federally Funded Innovation Inducement Prizes, Congressional Research Service. 7-5700 www.crs.gov R40677, June 29, 2009.

Zhang Y.-H.P., Lynd L.R. (2004), "Toward an aggregated understanding of enzymatic hydrolysis of cellulose: Non-complexed cellulase systems", Biotechnology and Bioengineering 88: 797-824.

Transitions

10

Abstract

The 21st century is an age of rapid change, principally because the Internet connects everyone to everyone else, ideas and money flow like contagions, innovation is progressing at an unparalleled pace, and the world is tilted. In this final chapter I offer these conclusions and observations: In the 21st century, *physical* society transitions to *virtual*, *local* transitions to *global*, and some *individuals* become as powerful as traditional authorities like government, law enforcement, and corporations. Political movements, dominant corporations, terrorist groups, online and physical flashmobs, newly formed nations, new ideas and fashions, scientific breakthroughs, and celebrity in the 21st century are products of Gause's Law of competitive exclusion. Nobody will be anonymous, and privacy will be considered old-fashioned—even anti-social. All of us will become increasingly dependent on a core set of sustainable commons, platforms, ecosystems, and networks that interact in complex ways. As such, they will be subject to self-organized criticality (SOC). Comparative advantage is the great equalizer of the 21st century. Humanity is likely facing a tipping point: we are terra-forming the planet from blue to orange. This transition will force huge global dislocations. Wealth inequality will destabilize entire nations and increase the likelihood of economic conflict as the tilted world seeks a new fixed point. Where that fixed point ends up is unknown. The 21st century optimizes on economic power rather than hegemony and military power. Winners will be nations that value innovators more than sports and movie celebrities, education over entertainment, risk-taking over entitlements, and "changing the world" over exploiting it.

10.1 One in a Million

If one person in a million wants to do harm, he or she has more power than ever before to wreck havoc on society. One part per million means over 7,000 perpetrators are empowered to hack the Internet, bomb the public square, take down a skyscraper, unleash a deadly pandemic, control a destructive airborne drone—even hold entire nations hostage to nuclear weapons as the leader of North Korea has demonstrated. The 21st century is not only the most exciting century mankind has ever experienced, but also the most dangerous.

The entire globe experiences the excitement of punctuated reality in unison—in time and space. We have built a rapid global collaboration communication network that is being used for both good and evil purposes. Vandals, revolutionaries, drug lords, racists, and anti-social malcontents now have the tools to marshal support for all kinds of movements, protests, dark and subversive activities that sway public opinion in both positive and negative directions. One influential person placed strategically in a social network—as a hub and/or an influential intermediary—can turn a random, unorganized herd into a focused social movement, as illustrated by the Occupy Wall Street flashmob. Online societies are especially vulnerable to persuasion when users lack conviction or awareness of the world they live in.

> The 21st century empowers both good and evil people to do incredible things. The question is, "How should society respond?"

First, social movements instigated by flashmob mentality are Levy Flights separated in time and space according to a long-tailed distribution. This means they happen in spurts and geographical proximity to one another. Like an epidemic, they have an epicenter that can only be detected at lightening speed by analyzing big data—social media and sentiment analysis. They happen so quickly that only real-time responses are adequate. Law enforcement authorities, product marketers, and trend-makers must fight fire with fire—they have to become an integral part of the self-organizing crowd. The implications for security, social stability, and product marketers are obvious.

In Chap. 2 we learned how simple and mindless flashmob self-organization is. Mobs of long-tailed size emerge from randomness with little provocation and a lot of peer pressure. The rule of group formation is incredibly trivial: the likelihood of an individual agreeing or disagreeing on any position is proportional to the positions of the individual's nearest (connected) neighbors. Polarization into groups (mobs) of a certain size obeys a long-tailed distribution—most groups are small, but some are large. The large ones can be troublesome for the establishment, as the Arab Spring illustrated.

Furthermore, the elapsed time between formations also obeys a long-tailed distribution. This means that most episodes are closely spaced together in time, but on rare occasion, episodes of significance occur only after a long pause. The calm before the storm does not mean everything is OK. It is simply a fact of punctuated reality.

As conviction of individuals increases, group polarization also increases. When individual conviction is strong, polarization no longer obeys a long-tailed distribution. Instead, two camps form—one "pro" and the other "con". The population becomes divided into camps much like the US Congress! Once again, this is nothing more than a random fluke of nature and yet more proof that social animals such as humans are mostly governed by randomness.

Flashmobs can be controlled and directed by *pinning* the most influential actors. The most influential actor is either the most connected or the one with the largest *betweeness* property. These "central" actors exert control over the mob proportional to their connectedness within the social network. The ultimate group outcome is also influenced by a handful of persuaders—neighbors that are listened to more than all others. These "loud" influencers can sway entire sub-networks in online communities such as Twitter.com.

These two properties—connectivity and "loudness" of neighbors—are levers to be used by marketers, politicians, and social activists. Entire communities can be turned into controlled flashmobs—both online and in the physical world through manipulation of these levers. Large flashmobs form when actors lacking conviction listen to a small number of strong influencers. On the contrary, large flashmobs are inhibited from forming when conviction is high or actors listen to more than a dozen neighbors. Flashmob intensity and frequency is amplified or attenuated by degree and strength of connectivity within a social network.

One person in a million may seem insignificant, but if that one-in-a-million person is highly connected, strong-willed, and persistent, he or she can sway an entire nation, social movement, political movement, or marketing campaign. A presidential candidate can be totally unknown in 2005, and become President of the USA in 2008. In the 21st century knowing the right social network buttons to push and network property levers to pull can magnify a one-in-a-million actor's power by a factor of a million.

Transition: In the 21st century, *physical* society transitions to *virtual*, *local* transitions to *global*, and some *individuals* become as powerful as traditional authorities like government, law enforcement, and corporations. The age of the non-governmental organization (NGO) has just begun and NGO power will continue to rise as the Internet is increasingly used to control public sentiment.

10.2 This Is Your Life

As I write this, billions of photos and tweets are being recorded on disk drives in the Internet cloud. Your sentiments, political leanings, buying habits, and current location are as public as if you bared your very soul in a 30-min spot on TV. And for the most part, we volunteer all this data about ourselves without a second thought. When combined, this data paints an accurate picture of who we are, where we are located, what were are about to do, and what we think. This continuous and unending global data roundup means the end of privacy and security as we know it. There is no more anonymity.

The 21st century can process kilo-zettabytes of data in a matter of hours—a feat that would have been completely out of the question in the 20th century. For example, 1,000 characters of personal identifiable information (PII) about every man, woman, and child on the planet can be stored on almost any personal computer. A server farm located somewhere in the cloud can hold millions of characters of PII on each one of us. Add nearly infinite processing power to this storage capacity, and any government, corporation, or curiosity-seeker can "look you up" in an instant and "read you like an open book".

Google.com passed one million servers on July 9, 2008, and reportedly surpassed two million in 2013. An estimate of the number of Amazon.com servers exceeds 500,000, and Facebook.com reportedly used 180,000 servers in 2012. These estimates are based on power consumption, because these companies do not report exact numbers. Technologists claim big data servers consume 1.5 % of the world's electrical power. It is no wonder that one of Google's data centers is located next to the largest power generator in the US—Bonneville Power Authority on the Columbia River in Oregon. These machines are culling through every tweet, Facebook.com post, Google gmail, and Pinterest.com photo online to get a "fix" on who you are, what you want, and what you are about to do next.

Every microsecond of every day, 360 days per year, these machines apply various *data mining* techniques to learn about your preferences ranging from products you buy, company you keep, money you earn, cars you drive, to your political biases and voting behavior. In fact, the most useful data mining software is a class of *machine learning algorithms* that literally learns everything it can about you. The cloud servers know you perhaps better than you know yourself. And what they do not already know, the machine learning algorithms can deduce.

Transition: In the 21st century nobody will be anonymous, and privacy will be considered old-fashioned—even anti-social. The lives of each of the 7.5–11 billion current and future citizens of earth will be recorded in such detail that governments, corporations, and "friends" will be able to replay anyone's life upon demand. In the 21st century everyone's life is literally an "open book."

10.3 Out of Nothing Comes Something

One of the repeated patterns of the modern era is the way new ideas, structures, innovations, conflicts, and solutions emerge from chaos rather than carefully laid plans. John Lennon once said, "Life is what happens to you while you are making plans!" And it is true: the personal computer, Internet, GPS/Navigation, robotic drones, global financial systems, new forms of government, and unprecedented wealth are products of unplanned emergence.

The foundations of modern society are mostly the result of accidental combinations of unlikely components. The personal computer emerged from the space program, as did GPS/navigation and the Internet. They were not planned. But they also were not random walks in the cosmos. They evolved according to a set of basic rules of nature. One of these rules—*Gause's Law*—governs the formation of major corporations and non-profit organizations across the globe. Gause's Law— also known as the *competitive exclusion principle*—explains how AT&T, Microsoft, and other corporations rose to monopolistic domination, and predicts how the global Internet is likely to evolve over the next 20 years.

The primary tool of competitive exclusion principle is *preferential attachment*. Here is how it works. At some point in the evolution of a market, company, idea, product, social group formation, or actor in general, one of the actors gains a slight advantage over all others. This advantage accelerates growth of the advantage itself, such that one actor becomes preferred over all others. This advantage grows logistically until all other actors are forced aside.

Gause's competitive exclusion principle is at work today in the form of Internet restructuring—the reshaping and domination of the global Internet by a handful of companies. As preferential attachment increases the advantage of one Internet provider over all others, the handful of companies that "own the Internet" decline in numbers. Eventually, one company will "own the Internet", unless governments intervene as the US government intervened to break up AT&T. So far, no single government has the power to stop the exclusion principle.

Preferential attachment and competitive exclusion are powerful forces that drive most emergent processes in the 21st century. Gause's Law can be observed in many parts of modern society. For example, dominant groups constantly emerge on Twitter.com and Facebook.com. Google.com is already a virtual monopoly in the Western World, and continues to spread its dominance around the globe. Global companies like Google.com and Facebook.com provide platforms for Gause's Law to repeat its fractal pattern of preferential attachment. For example, as Facebook rose to dominance, Zynga.com repeated the same pattern within Facebook. As the Apple iPhone rose to dominance, iTunes repeated the same pattern within the "Apple ecosystem". Corporations are smart to promote these niches within their walled gardens, because preferential attachment is a magnet for market share.

These global communication infrastructures constantly incubate new subsocieties as preferential attachment works its magic on people and organizations. Gause's Law is a propellant for self-similar fractals that create dominant organizations in the 21st century. It is just as valid today as it was 100 years ago when the first monopolies grew up with the new United States of America. In a sense, the 21st century is more like the 19th century than the 20th century, because of Gause's Law.

Transition: Political movements, dominant corporations, terrorist groups, online and physical flashmobs, newly formed nations, new ideas and fashions, scientific breakthroughs, and celebrity in the 21st century are products of Gause's Law of competitive exclusion. In an age of lightening fast global communications and powerful individuals, the competitive exclusion principle will become even more important. Success in almost every endeavor will require mastery of Gause's Law, but also its application as a self-similar fractal—repeating the process of turning small advantage into dominance of niches within niches.

10.4 The Commons

An *industrial commons*—or commons in general—is a resource shared by a community. For example, the Interstate Highway System, Fisheries, Forests, Global banking system, Social Security, Open Source Software, and Public Parks are well-known commons, because they encapsulate a resource shared by a community of users. A commons includes an infrastructure and the entities that depend on it. Other names for a commons are: platform, ecosystem, backbone, and infrastructure. The future of commerce and public-private partnerships lies in how well we understand these commons, and how well we manage them.

In a highly connected world, various commons are like islands—partially isolated but connected to one another. The global trade network is a commons that depends on the global financial network and global communications platform. And the financial network depends on the economies of nations, transportation networks, etc. Therefore, a collapse or enrichment on one commons affects the others. Industrial pollution of the air affects the "health care commons", and a shortage of food in one part of the world affects the price of grain in other parts.

A number of commons have become phenomenally successful, while others have become infamous failures. For example, Silicon Valley is known throughout the world as a very successful industrial commons. On the other hand, the communist form of government practiced by the Former Soviet Union (FSU) failed spectacularly as a commons—perhaps because it tried to share too much. [The FSU failure illustrates a *tragedy of the commons* as described in an earlier chapter]. The Interstate Highway System of the US was a phenomenal success as a shared commons, as was the first 40 years of the open Internet. Shared access to Earth's atmosphere remains a challenge—and we do not yet know if sharing the (polluted) skies is going to be successful or disastrous.

> My point is that some commons are "good" and some are not. Some improve life and some work against improvement. How do we know which is which?

The answer to this question is important because success in the 21st century depends on how the inhabitants of Earth manage various commons—both industrial and natural. The 20th century built a number of very successful commons—a global banking system, trade networks, and political commons such as NATO and the UN, etc. In the 21st century we will be challenged to create new and novel commons to solve wicked problems such as shortage of land, water, pure air, political and social justice, and avoid globally destructive wars. So, understanding how a sustainable commons operates is critical.

By 2013 the bottom-up process of self-organization was beginning to transform government through local and state governance. As the US Congress becomes deadlocked and incapable of compromise, major metropolitan and state governments have become the innovators and agents of change. States like Colorado and Washington have usurped the Federal government's control of marijuana, same-sex marriage, and health care. These trickle-up changes will emerge as global patterns and change America's politics.

No commons stands alone. In fact, every commons is actually an ecosystem where predators and prey roam and compete. The success of a commons, then, depends on the ecosystem's stability. Stable ecosystems survive, and unstable commons either become stable or die out. The key to a "good" versus "failed" commons is found in its *carrying capacity*. A "good" commons is one that is sustainable. [I purposely avoid ethical and moral arguments of "good" and "bad", because nature cares only for survivability, not morality]. Thus, the US Interstate Highway System is "good" because it has survived many decades and appears to be sustainable, while the communist system of the Former Soviet Union is "bad", because it was not sustainable. [Is the communist system of the People's Republic of China sustainable?]

Prosperity and success in the 21st century will depend on a number of critical commons. These are the existing and newly forming platforms for wealth generation—global energy networks, global communication networks, global trading networks, global governing institutions, and globally sustainable water, air, and food platforms. I am sure the reader can supply more examples of essential commons.

Sustainability can only be achieved by understanding the complex relationships between competitive forces operating in most commons—predator versus prey—and carrying capacity—the sustainability of an infrastructure in the face of demands placed on it. When these forces are balanced, the commons is stable and sustainable. When they are unbalanced, as illustrated by Minsky moments and excessive financial enrichment, the commons becomes instable and is likely to collapse. Unfortunately, collapse is not as rare as we think. The fact that collapses take us by surprise, suggests that we do not fully understand the dynamic behavior of essential commons.

The long-tailed consequences defined by self-similar, size-space-time continuum of punctuated reality suggests that considerable effort is required to sustain vital commons such as the global food supply, climate, trade, wealth, etc. This balancing act will become increasingly important and critical as resources are put under population pressure. We are at the beginning of a transition from fragmented individualism and rogue nationhood to a highly connected set of sustainable commons. These essential commons are economic *backbones*, technological *platforms*, and food, water, energy, and financial ecosystems that interact in non-linear (and often surprising) ways. To keep them alive and sustainable, we must understand complexity.

Transition: In the 21st century "no man is an Island". Instead, all of us will become increasingly dependent on a core set of sustainable commons, platforms, ecosystems, and networks that interact in complex ways. As such, they will be subject to self-organized criticality (SOC)—a property of complex systems that must be moderated because SOC increases and magnifies consequences. They are also vulnerable to the paradox of enrichment (POE) and tragedy of the commons (TOC). Therefore, a basic responsibility of governance is to promote sustainable, resilient, and robust commons, platforms, and ecosystems. [These commons are global and not controlled by one single nation. They are also subject to cascades and unexpected shocks as illustrated by the Fukushima dia-ichi nuclear power plant meltdown].

10.5 The Wealth of Nations

Comparative advantage is making China the largest economy in the world. It is the economic force that drives national prosperity, and indirectly, global power. By trading what a country does well for goods and services that it does not do well, any country can turn itself into an economic powerhouse. China is the current exemplar and India is right behind it. The correlation between trade and GDP is too strong to ignore—big traders will dominate the globe in the 21st century.

Nearly two-thirds of the world has excess labor, while two-thirds of the wealth is in the pocketbooks of the Western nations that benefitted from the Industrial Revolution. Wealth of nations follows a long-tailed distribution just as wealth of individuals does. That is, wealth is a self-similar fractal—the long-tailed pattern that describes countries repeats as a long-tailed pattern within each country. But comparative advantage can be a leveling force of nature if exploited.

China and India have cheap labor—lots of it. The West has money—lots of it. The West is also hungry for products at low prices. This is a marriage made in comparative advantage heaven. It is also the secret to how over-populated poor countries can go from third world to first-world status. China and India may be the first to execute this strategy, but other nations are following. Soon, African and Islamic nations will follow.

The excessive spending and deep debt accumulated by Europe and the US over the past twenty years only makes it more likely that wealth will shift from West to East, North to South, sparsely populated to over populated, and old industrialized to new modernized countries. In fact, this shift is one of the most profound transitions of the 21st century.

The Western nations militarily dominated the world since World War Two. Now the challenge is to economically dominate the world through commerce. The West and North have no choice but to adapt to a comparative advantage driven world. They have only a handful of tools: education/innovation, immigration, and distributed manufacturing. They lack labor and natural resources.

Innovation through education and training will fuel Western-style comparative advantage (innovation is indeed an advantage). Innovative companies like *Google.com* and innovative products like the *Apple iPad* drive Western dominance even in the face of cheap labor and cheap resources. Immigration of highly educated people to the West fuels innovation, and when that is not enough, Western companies must distribute manufacturing and services to where the consumers are—in the East and South.

10.5 The Wealth of Nations

Because trade and economic power go hand-in-hand, the great trading countries like the US, UK, Germany, Netherlands, Sweden and Norway will continue to be powerhouses. But their dominance will require a shift from labor-intensive to mind-intensive industries. Also, these traders must continue to develop their trade networks—extending the EU further East and South, and extending NAFTA to the South. A political union similar to the EU may take root and emerge in the Western Hemisphere to bolster the USA-Canada-Mexico market against the onslaught of resource-hungry Asia. Similar alliances are likely to form in the second half of the century among African and Islamic nations.

> The world is tilted—labor in the East and South and capital in the West and North. What does a tilted world mean?

Transition: Comparative advantage is the great equalizer of the 21st century. It will draw wealth from West to East, and North to South. Furthermore, global power equals economic power, meaning that trader nations will dominate in this century. First, the highly trained work force countries, and then the highly populated and cheap labor countries will siphon wealth from the West. The currency of exchange will be cheap labor and expensive innovative know-how. The elite of the over populated countries will migrate to the West because of its high living standard, and the highly industrialized Western nations will distribute manufacturing and services to densely populated regions of the world—to where vast numbers of consumers live. Western nations may succeed in integrating their trading (commons) platforms and trading networks to fend off the growing power and influence of the East and South, but if not, they will decline in power and importance.

10.6 Extreme Challenges

Fortunately, the easy problems of the world have largely been solved. Poverty, disease, war, hunger have been avoided or are avoidable if we set our minds to them. The 20th century was very successful in terms of improved living standards, safety, and security of billions of people. The future is bright. Or is it?

The next generation of challenges facing mankind is breathtaking. It is highly likely that mankind will face extremes never faced before. At the low end are superstorms ten times bigger than any we have seen thus far. Natural disasters are on the increase and will continue to grow until and unless global climate change is reversed. But even if we repeal climate warming, the possibility of annihilation from a collision with an errant asteroid, massive death from a genetically engineered virus, or collapse of civilized society because of poor policy decisions looms large. The challenges lying ahead are much more complex than ever before. They require a deeper understanding of self-organization, tragedy of the commons, paradox of enrichment, competitive exclusion principles, and risk assessment.

Current leadership in government knows almost nothing about these complexity theory concepts. Even less is understood by our leaders of how to apply complexity theory principles to problem solving. Unfortunately, the tide is turning away from deep thoughtful consideration of unintended consequences, non-linear behavior, and highly interconnected conditional probabilistic reality towards bickering, greed, and avarice. Our leaders simple don't have a clue, and they are not motivated to get one. This is leading us to the edge of disaster.

Transition: Humanity is likely facing a tipping point: we are terra-forming the planet from blue to orange. This transition will force huge global dislocations—populations receding from coastlines, movement of vast supplies of potable water from the sea to irrigate parched land, gigantic investments in weather-related disasters, and major shifts in the world's supply of food. The demand for natural resources likely will lead to conflict and political shifts that redraw borders, political alliances, and reshape national strategies of the world's leading economies. Only a better understanding of complexity theory will prevent politicians from making disastrous policy decisions.

10.7 The World May Be Flat, but It Is Also Tipped

Mass starvation did not wipe out humanity in the 1970s and 1980s as predicted, but something else almost as earthshaking happened, instead. Humans doubled in numbers and the world tipped—most of the wealth piled up in Europe and North America, and most of the population stacked up in Asia and Africa. Money slid North and West; population slid South and East. People are moving from South to North and East to West.

Tilted world dynamics will play out even more dramatically over the next half-century, because concentrated capital in one part of the world will seek out cheap labor in another part of the world. Counterbalancing this is the rapid rise of consumer markets in the East and South that will attract *distributed manufacturing*—the distribution of production facilities to where the customers are. Commercial airplanes, automobiles, pharmaceutical drugs, electronic devices, and numerous service industries will be manufactured across the globe with most of the action near buyers.

While manufacturing is being distributed to where products are consumed, financial control continues to concentrate in the West and North. Trade networks distribute production and goods, but money flows are subject to *scale-free network* effects. A mere 737 financial firms have accumulated 80 % of the control over the value of all transnational corporations on the planet! "Nearly 40 % of the control over the economic value of transnational corporations in the world is held, via a complicated web of ownership relations, by a group of 147 transnational corporations in the core, which has almost full control of itself," say Vitali, Glattfelder, and Battiston, researchers at ETH-Zürich—Eidgenössische Technische Hochschule, Zurich, Switzerland.

The top 10 financial powerhouses on the planet in 2005 were located in the West and North: Barclays PLC, AXA, State Street Corp, JPMorgan Chase, Legal & General Group, UBS AG, Merrill Lynch, Deutsche Bank, Franklin Resources, and Credit Suisse Group. Money flows through the global communications networks, but goods and services have to be transported. This may explain how and why the West and North is able to maintain control of wealth without "being there."

Wealth has always been distributed according to a long-tailed Pareto curve. Most people in the world are poor or middle-classed. Relatively small groups of people are rich, and only handfuls are super-rich. This is true of nations, corporations, and individuals. [Self-similar fractals, again!] We find Pareto's curve at just about every scale of humanity. Why?

Wealth increase is proportional to the amount of wealth already owned.

This is simply another way of stating the *law of increasing returns*—a type of preferential attachment. This powerful force of nature will continue to shape the wealth of individuals and nations unless disrupted by governments. It is as true in the 21st century as it was in all previous centuries.

Tilted world economics, Pareto distributed wealth, and massive shifts in manufacturing toward consumers will inevitably destabilize societies around the globe. These nonlinearities pose challenges to social and political stability. And the tectonics will be so pronounced that nobody will be immune to the tremors. A tilted world is an unstable world.

Transition: Wealth in the 21st century will accumulate according to the Pareto distribution just as it has in the 20th century. But wealth redistribution will have to deal with the shift of West and North dominance to East and South rising economic power. This transition will destabilize entire nations and increase the likelihood of economic conflict as the tilted world seeks a new fixed point. Where that fixed point ends up is unknown.

10.8 The Millennium Falcon

Twelve astronauts visited the Moon between 1969 and 1972. Then nothing. Exploration of space essentially stalled almost as soon as it started. NASA's budget peaked in 1963 at $33 billion (in inflation adjusted dollars), dropped precipitously to $11 billion by 1973, and flattened out to $19 billion by 2013. The world became obsessed with conflict, globalization, and protection of the environment. Space travel lost its luster, and people lost faith that humans would ever be able to live on the Moon or populate Mars, according to Elon Musk, entrepreneur, innovator, and billionaire CEO of SpaceX, the first commercial company to put a spaceship into orbit.

Musk made his first fortune at age 24 from the sale of PayPal to eBay. Then he quickly moved on to reinvigorate space travel by forming SpaceX—a Hawthorne, California company with a $1.6 billion contract from NASA to fly a dozen cargo missions between earth and the International Space Station. SpaceX radically reduced the cost of boosting cargo into space by using "off-the-shelf" parts (Musk bought one component from eBay). His thesis is that people need to believe that space travel is possible before they will get excited about it again like they did in the 1960s. The SpaceX workhorse booster is named the *Falcon*, after the *Millennium Falcon* starship in the iconic movie *Star Wars*.

Musk is often considered the inspiration for *Tony Stark*, the industrialist lead in the movie *Iron Man*. Indeed, Musk had a cameo role in *Iron Man 2*. But Musk is bigger in life than Tony Stark, because he didn't stop with SpaceX. He invested in his cousin's company SolarCity, to transform the energy sector, and served as its CEO while also building the first commercially successful electric car company, *Tesla Motors*. Tony Stark is a one-trick pony—Elon Musk is on his fourth groundbreaking company. And he turned 40 in 2012.

Musk may be the next Steve Jobs, but unlike Jobs, Elon Musk lives in the 21st century where economic power has supplanted hegemony as a national strategy, and online virtual communities have replaced the public square. Countries like Brazil, Russia, India, and China (BRICS), are pursuing economic power, not military power. Instead of building weapons of mass destruction and spreading ideology, the BRICS are pursuing higher living standards and economic power. This changes everything.

Leading 21st century countries realize national security and global power derive from economic—not military power. They are focused on leaps in space exploration, electric transportation, clean energy, bioengineering, materials science, nanotechnology, and other breakthrough technologies that promise to radically improve the economic wellbeing of their people. They have small or non-existent military budgets and very large education, research, and economic development budgets. They invest in the 50-year horizon rather than the 3-month horizon.

Future success of nations depends on the cultivation of more people like Elon Musk. And the door is open for the type of human exploration proposed by Musk, because leaps increase economic power while violent conflicts sap economic strength. So the 21st century is an optimistic century. But only if we value risk over entitlements, hard work over entertainment, and trade over hegemony.

Transition: The 21st century optimizes on economic power rather than hegemony and military power. The pace of innovation will exceed that of the 20th century (the long tail distribution of innovation is shorter). Winners will be nations that value innovators more than sports and movie celebrities, education over entertainment, risk-taking over entitlements, and "changing the world" over exploiting it.

The extremes of the 21st century promise to be bigger, more shocking, more destructive, and more dramatic than the 20th century. But they also promise to be more hopeful, because leaps in technology—the long-tailed distribution of invention and innovation. From robotics, nanotechnology, breakthroughs in energy and terraforming earth, to life-extending advances in medicine, the 21st century extremes will be both dangerous and exciting. And laden with opportunity.

About the Author

Ted G. Lewis is an author, speaker, and consultant with expertise in applied complexity theory, homeland security, infrastructure systems, and early-stage startup strategies. He has served in both government, industry, and academe over a long career, including, Executive Director and Professor of Computer Science, Center for Homeland Defense and Security, Naval Postgraduate School, Monterey, CA. 93943, Naval Postgraduate School, Monterey, CA., Senior Vice President of Eastman Kodak, President and CEO of DaimlerChrysler Research and Technology, North America, Inc., and Professor of Computer Science at Oregon State University, Corvallis, OR. In addition, he has served as the Editor-in-Chief of a number of periodicals: IEEE Computer Magazine, IEEE Software Magazine, as a member of the IEEE Computer Society Board of Governors, and is currently Advisory Board Member of ACM Ubiquity and Cosmos+Taxis Journal (The Sociology of Hayek). He has published more than 30 books, most recently including Book of Extremes: The Complexity of Everyday Things, Bak's Sand Pile: Strategies for a Catastrophic World, Network Science: Theory and Practice, and Critical Infrastructure Protection in Homeland Security: Defending a Networked Nation. Lewis has authored or co-authored numerous scholarly articles in cross-disciplinary journals such as Cognitive Systems Research, Homeland Security Affairs Journal, Journal of Risk Finance, Journal of Information Warfare, and IEEE Parallel and Distributed Technology. Lewis resides with his wife and dog, in Monterey, California.

Index

A
Abbottabad, Pakistan, 35
Absolute zero, 110, 111
Actor, 48, 59, 62, 64, 167, 169
Albedo, 111
Algotrek Technology Consulting, 35
Allen, Paul, 152
Alpha Centauri, 150, 151
American Tea Party, 6, 60
Ansari, Amir, 151
Apache, 38
Apophis, 105, 106
A posteriori, 9, 122
A priori, 13, 42
Arab Spring, 6, 60, 61, 65, 166
Archon Genomics X-Prize, 152
ARkStorm, 129
Asian flu crisis, 96
Asilomar State Beach, 1
Autonomous system network (ASN), 58
Association of computing machinery, 45
Astronaut glove challenge, 153
Astroturfing, 41, 45
Athar, Sohaib, 35, 38
Atlantic conveyor, 107, 108
Atmospheric river, 128, 129
AT&T, 52–54, 57, 58, 66, 152, 162, 169
Autonomous system, 51, 58
Avery, Dennis T., 108

B
Baby bells, 53
Bak's Sand Pile, 36, 150
Barringer Crater, 106
Bass, Frank, 56, 93
Battiston, Stefano, 137
B'Austin Ale, 39
Bayesian network, 43, 45, 46
Bayes, Thomas, 42

Bayes' theorem, 42, 43, 45
Bean machine, 15, 16
Bell, Alexander Graham, 51, 52
Benner, Lance, 106
Big data, 14, 36–38, 40, 41, 43, 44, 46, 60, 144, 166, 168
Billion-dollar storms, 122
Binomial distribution, 15, 16
Black body, 110–112
Black plague, 132
Black swans, 120, 122, 129
Boguna, Marian, 98
Bond events, 108, 119
Bonneville Power Authority, 168
Borlaug, Norman, 132
Bot, 41
Bouchard, Jean-Philippe, 145
Boulevard Animali, 21
Boulevardiering, 21, 22, 26
Branson, Richard, 152
Bretton Woods Accord, 90
Bright tomorrow, 153
Burkle, Winifred, 94
Bursts, 2–5, 7–9, 11, 95
Busch, Michael W., 106
Butterfly parade, 8

C
Carboniferous rainforest collapse (CRC), 106
Carlton, Tamara, 157, 160
Carnegie, Andrew, 51
Carrington event, 108, 109, 110
Carrington, Richard, 108
Carrying capacity, 94, 171
Cars, 7–9, 11, 17, 46, 155, 156, 158, 168
Chabot Space and Science Center, 153
Chaos theory, 91–93
Chaotic adaptation, 6
Chernobyl, Ukraine, 9

Chinazzi, Matteo, 100
Chinese police academy, 33
Christensen, Clayton, 161
Churchman, Charles, 154
Cleveland Heights, Ohio, 22
Cloud-computing, 57
Cogent/PSI, 53, 58
Commons, 165, 170
Comparative advantage, 87–91, 96, 103, 136, 165, 172, 173
Competitive exclusion principle, 51, 53, 54, 56, 66, 67, 169, 173
Concentration of capital, 136
Conditional probability, 14, 16, 44
Corn Law, 89
Crowdsensing, 38–40, 44
Cutting, Doug, 38, 42

D
Dallas, Texas, 21, 56
Dansgaard-Oeschger (D-O) events, 108
DARPA Grand Challenge, 46
Data mining, 168
Defense Advanced Research Projects Agency (DARPA), 150
Degree of belief, 42, 44
Deutscher Commercial Internet Exchange, 58
Diamandis, Peter, 152
Diameter of network, 51, 63
Diffusion, 56, 57, 62, 93
Diffusion of innovation, 93
Digital Maoism, 49
Diminishing returns, 88, 89, 92, 136
Disintermediation, 52
Distributed production, 136
Distribution, long-tailed, 2–7, 11, 12, 16, 17, 23, 27, 30, 31, 58, 60, 62, 63, 103, 121, 131, 144–146, 166, 172
Distribution, normal, 2, 7, 8, 11, 15–18, 30, 144
Domain name server (DNS), 65
Dorothy Jemison Foundation for Excellence, 150
Dreamliner 787, 137
Dunbar, Robin Ian, 65
Dunbar's number, 65
Dyson, Freeman, 117

E
Early adopter, 93
Earthquakes, 6
Eastman Kodak, 161

Economic backbone, 171
Egypt, 6, 60, 104
Ehrlich, Paul, 131
Eisenhower, Dwight D., 79
El Chichon eruption, 119
Ellison, Larry, 1
Elsner, James, 121
Emissivity, 111
Energy budget, 110–112, 119–121
Epistemology, 40
Exceedence probability, 7, 8, 130, 120–122, 149, 150

F
Fagiolo, Giorgio, 100
False positive, 42, 43
Farmville, 48
Federal AID Highway Act of 1956, 79
Federal reserve, 38
Feigenbaum, Mitchell, 93, 94
Feigenbaum number, 94
Financial collapse, 6, 17
Financial contagion, 99, 100, 139
Firehose, 36, 37, 39, 44
First mover, 93
Fixed point, 111, 155, 162, 165, 175
Flashmob, 6, 166, 167
Fold.it, 48, 49
Former Soviet Union, 105, 170, 171
Fractal, 5, 6, 8, 9, 11, 12, 16, 17, 19, 115, 129, 144, 145, 147, 148, 169, 172
Fractal dimension, 6, 9, 19, 25, 144, 145, 147, 148
Free trade, 89, 90, 100
Freedom prize, 153
Friedman, Thomas, 135
Fukushima dai-ichi, 8, 9, 12, 14, 17

G
Galton, Sir Francis, 14
Gambling, 12
Gamification, 47
Gause, Georgii Frantsevich, 53
Gross domestic product (GDP), 72, 74, 75, 88, 90, 93, 94, 96, 98–100, 104, 113, 120, 172
Geomagnetic storm, 109
Gerald J. Ford Stadium, 22
Gini, Corrado, 144
Gini index, 144, 145
Giorgini, Jon, 106
Glattfelder, James, 137, 138, 174
Global 500 corporations, 140, 143

Index

Godel's theorem, 47
Gompertz, Benjamin, 93
Good, Jonathan, 161
Google Lunar X-Prize, 152
Graeber, David, 26
Gray body, 111
Great East Japan earthquake, 8, 11, 102
Green paper, 65, 66
Green revolution, 32
Green's function, 117, 118
Global top-level domain (GTLD), 65
Guangdong, China, 33

H
Hadoop, 38, 39
HADSST2 data, 112, 114, 116
Haining, Zhejiang, 33
Halting problem, 46, 47
Hansen, James Edward, 117
Hansen–Sato function, 117, 119
Hardin, Garrett, 77, 79
Hashtag, 60, 62
Hayden, Tom, 26
Heinrich events, 108
Highway Trust Fund, 80
Hodges, Robert, 121, 122
Homeownership, 69, 71–76
Homophily, 63, 64
Housing bubble of 2001–2006, 70
Hub, 40, 55–58, 60, 61, 63–65, 67, 101, 139, 166
Hubbard, Gardiner, 51
Hubbert, Marion King, 93
Hurricanes, 6, 11, 17, 121, 122
Hydrogen prize, 153

I
IANA, 26
Internet corporation for assigned names and numbers (ICANN), 66, 67
Incentivized competition, 152, 153
Increasing returns, 54
Industrial revolution, 42, 132, 172
Innovation butterflies, 155
Innovator's dilemma, 161, 162
Internet animali, 21–23, 27, 32
Interstate highway system, 170, 171
Intergovernmental panel on climate change (IPCC), 118–121
Irvington, New Jersey, 23

J
Jackson, Mississippi, 23
Japan's Bubble Economy of 1980s, 70
Jemison, Mae, 149, 150, 153, 162
Jobs, Steve, 1, 52, 176
Jet propulsion lab (JPL), 105

K
Keith, Robert, 101
Kelly, Kevin, 49
Keynes, John Maynard, 88
Kingsbury commitment, 52
Krakatau eruption, 119
Kurzweil, Raymond, 49

L
Lake Agassiz, 107
Lake Victoria, 40
Lanier, Jaron Zepel, 49
Lasn, Kallie, 25, 26, 31
Law of increasing returns, 175
Lennon, John, 168
Level-3 communications, 53
Levy flight, 17, 19, 45, 149, 166
Levy walk, 8, 9, 11, 16, 128
Lifecycle, 93
Lindberg, Charles, 152
Lindsey, Rebecca, 119, 130
Liping, Sun, 33
Little Ice Age, 108
Lloyd, William Forster, 77
Logistical growth function, 92–94
Long-tailed distribution, 2–6, 9, 15–17, 19, 103, 128, 145–147, 166, 167
Lynd, Lee, 155, 156

M
Machine learning, 45, 46, 168
Malaria, 40
Malthus, Thomas, 132
MapReduce, 38, 44
Master of Chaos, 94
McCulley, Paul, 70
Media Guardian Innovation Award, 38
Meme, 65, 71
Menczer, Filippo, 41
Meridional overturning circulation (MOC), 107
Meta-stable, 92

Mezard, Marc, 145
Middle East, 14, 67, 135
Milankovich, Milutin, 108
Milgram, Stanley, 64
Milwaukee, Wisconsin, 22
Minsky, Hyman, 70
Minsky moment, 69–71, 171
Mississippi Bubble of 1719–1720, 70
Mojave Aerospace Ventures, 152
Mongolia, 33
Monte Carlo, 154
Monterey Bay Aquarium, 1
Moral hazard, 82, 83
Morgan, Julia, 1
Musk, Elon, 175, 176

N
Nairobi, Kenya, 40
Napoleonic War, 88
National System of Interstate and Defense Highways, 79
Natural monopoly, 52
Naval Postgraduate School, 48, 159
Negative forcing, 118–120, 122
Nelson, Admiral, 88
Net neutrality, 66
Network effect, 34, 54, 143, 174
Network science, 140
Networked, 12, 63
New Yorkers against budget cuts, 26
Non-governmental organization (NGO), 101, 167
Nieto-Gomez, Rodrigo, 47
Not in my back yard (NIMBY), 160
Ninety-nine percent, 26
Nokia sensing X CHALLENGE, 153
Nolan, Michael C., 106
Normal distribution, 1, 2, 4, 7, 8, 11, 16, 17, 31
Northrop Grumman Lunar Lander X CHALLENGE, 152
National Telecommunications and Information Administration (NTIA), 65
Nuclear power, 6, 9–11, 14, 82, 83, 102, 172

O
Occupy wall street, 6, 23, 25, 26, 266
One percent, Irish hunger strike, 26
Open source software, 76, 170
Orteig prize, 152
Orteig, Raymond, 152
Osama bin Laden, 35–37
Ostro, Steven, 106

Outlier, 2, 3, 40, 42

P
Pacific Grove, 1
Pagerank, 62
Pandemic, 14, 166
Paradox of enrichment, 69, 72, 73, 76, 82, 87, 88, 93, 144, 148, 171, 173
Pareto index, 144
Pareto, Vilfredo Federico Damaso, 144
Pascal, Blaise, 12
Pearl, Judea, 45
Pearl, Raymond, 93
Peering, 58
Permian–Triassic extinction (Great Dying), 106
Philadelphia, Pennsylvania, 26
Physical layer internet, 57
Pinatubo eruption, 119
Pineapple express, 122, 127, 129
Pinning, 31, 32, 167
Poisson distribution, 7
Pollock, Jackson, 114
Population bomb, 131, 133
Predator, 58, 72, 74, 77, 80, 171
Predator-prey, 72, 77, 80, 81, 84
Predictive analytics, 14, 39
Preferential attachment, 51, 53–56, 58–60, 63, 67, 131, 138, 140–142, 146, 169, 175
Prey, 72, 74, 75, 77, 171
Price-Anderson Nuclear Industries Indemnity Act of 1957, 82
Price, Richard, 42
Progressive automotive prize, 153
Progressive Insurance Automotive X-Prize, 152
Prometheus, 155, 156
Public parks, 76, 170
Punctuated, 1, 3, 4, 7, 8, 12, 18, 21, 23, 52, 69, 93, 105, 166, 171
Punctuated reality, 5, 17, 166

Q
Qualcomm Tricorder X-Prize, 153
Qwest, 53, 58, 152

R
80–20 % rule, 144, 148
ReallyVirtual, 35
Reed, Lowell, 93
Regression to the mean, 16

Retweet, 60
Reyes, Javier, 100
Ricardo, David, 88
Roaring twenties of 1924–1929, 70
Robin Hood tax, 26
Rockerfeller, John D., 51
Rogers, Everett M., 93
Rootkit, 156, 157, 162
Rosenzweig, Michael, 72
Runaway greenhouse effect, 117
Russian financial crisis of 1998, 70
Russian virus, 96
Rutan, Burt, 152

S
Salad bowl of America, 159
Sanders, Thomas, 51
SARS, 12, 27
Scalable, 5
Scale, 4, 5, 9, 60, 63, 67, 110, 115, 120, 138, 139, 147, 175
Scale-free structure, 60, 65
Schiavo, Stefano, 99
Schwartz, Mattathias, 26
Science of social, 36
Self-organization, 21, 31, 32, 51, 56, 57, 61, 131, 143, 166, 170, 173
Self-similar, 5, 6, 9, 71, 115, 129, 169, 171, 175
Sense-making, 35, 38, 40, 43
Serrano, Angeles, 98
Silver Springs, Maryland, 22
Singer, Fred S., 108
Small world, 51, 63, 64, 98
Smith, Adam, 87
Social network, 25, 27, 29–31, 34, 51, 59–61, 63, 64, 167, 155
South Sea Bubble of 1720, 70
Southern Methodist University, 21
Space shuttle, 150
Spectral radius, 31, 32
Spirit of St. Louis, 152
Spontaneous order, 27, 53
Sea surface temperature (SST), 108, 112, 113, 115, 121
STANCE, 48
Standard Oil Corporation, 51
Stanley, 46, 64
Star Trek: The Next Generation, 49, 150
State-space diagram, 73–76, 80, 81, 114, 115
Stefan–Boltzmann Law, 110, 111
Strongly connected component (SCC), 138–142

Subprime mortgage, 69, 75
Sunset drive, 7, 8
Superstorm, 108, 120, 122, 173

T
2001 Tohoku earthquake, 8
Telecom hotels, 57
Telecommunications Act of 1934, 52
Telecommunications Act of 1996, 52, 66
Tequila crisis, 96
Terrorists, 48
Twenty-foot Equivalent Unit (TEU), 101
Thermohaline circulation, 107, 118
Thomson, Sir William—Lord Kelvin of Scotland, 110
Three-mile island, 9, 82
Throttling, 66
Thrun, Sebastian, 46
Tier-1 Internet Service Providers, 53, 66
T-Mobile, 53
Topsy.com, 39, 60
Tourists, 7, 152, 159
Toyota Motors, 11
Truthy, 41
Tsar bomb, 105
Tsinghua University, 33
Tulip mania of 1637, 69
Tunisia, 6, 12, 25
Tunney, Justine, 26
Turing, Alan, 47
Turing award, 45
Tweet, 28, 35, 37, 41, 60, 63, 64, 168

U
Unconstrained optimization, 154, 155, 162
United Nations Internet Governance Forum, 67
University of Mississippi, 21

V
Vailism, 52
Vail, Theodore, 52
Vanderbilt, Cornelius, 51
Verhulst, Pierre Francois, 92
Verizon, 53, 162
Vertical integration, 52
Virgin SpaceShip, 152
Vitali, Stefania, 137
Vodafone, 53

W
Walled garden, 52, 169
Wal-Mart, 22, 39, 143
Watson, Reverend Henry William, 15
Wealth effect, 132, 136
Weir, Jeffrey H., 159
Wendy Schmidt Oil Cleanup X CHAL-
  LENGE, 153
Western electric, 52
White space, 149, 160, 162
White, Micah, 25
Wicked problem, 153–155, 170
Wilson, Woodrow, 52
Wolf prize, 94
World trade web, 12, 87, 96, 97, 102, 103
Wukan, China, 33
Wu, Michael, 36
Wu, Tim, 66

X
Xian, S., 116
X-Prize foundation, 152

Y
100 year starship (100YSS), 150
Younger Dryas stadial—the Big Freeze, 107

Z
Zhang, Percival Y-HP., 155
Zhen-Shen, L., 101

Levy Walk/Flight (as opposed to random walk)